Anton Malygin

Design and Experimental Investigation of a Second Harmonic 20 kW Class 28 GHz Gyrotron for Evaluation of New Emitter Technologies

Karlsruher Forschungsberichte aus dem
Institut für Hochleistungsimpuls- und Mikrowellentechnik

Herausgeber: Prof. Dr.-Ing. John Jelonnek

Band 10

Design and Experimental Investigation of a Second Harmonic 20 kW Class 28 GHz Gyrotron for Evaluation of New Emitter Technologies

by
Anton Malygin

KIT Scientific Publishing

Dissertation, Karlsruher Institut für Technologie (KIT)
Fakultät für Elektrotechnik und Informationstechnik, 2016
Referenten: Prof. Dr. rer. nat. Dr. h.c. Manfred Thumm
Prof. Dr.-Ing. Arne F. Jacob

Impressum

KIT Scientific Publishing

Karlsruher Institut für Technologie (KIT)
KIT Scientific Publishing
Straße am Forum 2
D-76131 Karlsruhe

www.ksp.kit.edu

Print on Demand 2016

ISSN 2192-2764
ISBN 978-3-7315-0584-6
DOI: 10.5445/KSP/1000059748

Forward of the Editor

New approaches for sustainable methods of environmentally friendly energy generation are required to support the rapid growth of the world population and economy. One future candidate is the controlled thermonuclear fusion in magnetically confined plasmas where high-power microwave sources (gyrotrons) are used for plasma heating and stabilization to achieve the required temperatures of around 120 Million degree Celsius. Gyrotrons are electron cyclotron masers driven by weakly relativistic helical electron beams, operating in a longitudinal magnetic field and highly oversized cavities. The current state-of-the-art for frequencies up to 170 GHz is 1 MW output power in continuous wave (CW) operation at 50 % efficiency, including single-stage depressed collector operation. Future gyrotrons shall be able to generate output powers above 2 MW with higher than 60 % total efficiency at operating frequencies above 200 GHz. To achieve those targets, new emitter technologies are required. Appropriate test devices are the base for that developments. Additionally, industrial microwave processes are demanding for the possibility to use efficient medium power gyrotrons with output powers at around 20 kW CW at industrial frequencies up to 30 GHz.

In the present work, Dr.-Ing. Anton Malygin combines both demands in an ideal way. He designed a 22 kW CW, 28 GHz gyrotron with a world record efficiency of 43 % (without energy recuperation), operating at the 2nd harmonic of the electron cyclotron frequency. The chosen parameters for the emitter current density allow the direct evaluation of new emitter technologies for future highly efficient and reliable fusion gyrotrons. It is supported by the unique modular concept of the gyrotron. The chosen operating frequency, achieved output power and efficiency allow the direct application and scaling to the industrial ISM band at 24 GHz for material processing applications at the Institute for Pulsed Power and Microwave Technology (IHM) at Karlsruhe Institute of Technology (KIT).

Dr.-Ing. Anton Malygin is providing the fusion community and the industrial community using gyrotrons for electron cyclotron resonance heating and materials processing a very powerful tool for emitter testing and direct industrial use.

Design and Experimental Investigation of a Second Harmonic 20 kW Class 28 GHz Gyrotron for Evaluation of New Emitter Technologies

Zur Erlangung des akademischen Grades eines

DOKTOR-INGENIEURS

von der Fakultät für
Elektrotechnik und Informationstechnik
des Karlsruher Institut für Technologie (KIT)

genehmigte

DISSERTATION

von

Dipl.-Ing. Anton Malygin
geb. in Nischni Nowgorod, Russland

Tag der mündlichen Prüfung:	27.07.2016
Hauptreferent:	Prof. Dr. rer. nat. Dr. h.c. Manfred Thumm
Korreferent:	Prof. Dr.-Ing. Arne F. Jacob

Kurzfassung

In dieser Arbeit wird über die Auslegung und experimentelle Untersuchung eines bei der 2. Harmonischen der Elektron-Zyklotron-Frequenz arbeitenden 20 kW / 28 GHz Gyrotrons berichtet. Dieses Gyrotron wurde so ausgelegt, dass neue Emitter-Technologien für zukünftige Hochleistungsfusionsgyrotrons untersucht werden können. Um solche neuartigen Emitter-Technologien erforschen zu können, muss die Auslegung des Gyrotrons mehrere Anforderungen erfüllen. Erstens, muss die Stromdichte an der Emitter-Oberfläche ähnlich groß wie bei Fusionsgyrotrons der Megawattklasse sein. Zweitens, muss das Gyrotron wie Fusionsgyrotrons im kontinuierlichen Betrieb (continuous wave: CW) arbeiten. Drittens, soll das Gyrotron in einem weiten Bereich von Parametern des Elektronenstrahls betrieben werden können, um die Betriebseigenschaften des Emitters in diesen verschiedenen Betriebsregimen untersuchen zu können. Und schließlich viertens, muss das Gyrotron so gestal-tet sein, dass es die Möglichkeiten bietet, zur Anwendung bei der Materialprozesstechnik im IHM eingesetzt werden zu können.

In einem ersten Schritt, wird die Auslegung des Resonators ge-zeigt. Um stabilen Einmodenbetrieb zu erzielen, wurden Simulati-onsrechnungen mit einem nichtstationären Vielmoden-Code durchgeführt. Die Optimierung des Resonatorprofils ergab die Unterdrückung von Störmoden und damit den stabilen Betrieb im Hauptmodus bei der zweiten Harmonischen der Elektron-Zyklotron-Frequenz. Es wurden Simulationsrechnungen zur Trioden-Elektronenkanone durchgeführt, um eine gute Qualität des Elektronenstrahls in einem weiten Bereich von Betriebspara-metern zu erreichen. Eine Elektronenkanone vom Triodentyp wurde verwendet, um den Bereich der Betriebsparameter zu vergrößern. Das Profil des Gyrotron-Kollektors wurde so optimiert, dass die thermische Belastung der inneren Oberfläche reduziert wird, und somit CW-Betrieb möglich ist.

Die experimentellen Untersuchungen zeigten stabilen CW-Betrieb des Gyrotrons im gewünschten Hauptmodus $TE_{1,2}$. Die Trioden-Elektronenkanone und der Resonator arbeiteten stabil in einem weiten Bereich von Betriebsparametern und erlaubten, einen hohen Wirkungsgrad im Hauptmodus zu erhalten. Der Vergleich der experimentellen und theoretischen Ergebnisse zeigte sehr gute Übereinstimmung.

Die folgenden experimentellen Ergebnisse wurden erzielt: 22.5 kW Mikrowellenausgangsleistung bei 23.4 kV Elektronen-strahlspannung und 2.23 A Elektronenstrahlstrom mit einem Weltrekordwirkungsgrad für bei der 2. Harmonischen der Zyklotron-Frequenz arbeitenden 28 GHz CW-Gyrotrons von 43 %.

Abstract

In this work, the design and experimental investigation of a 20 kW/ 28 GHz gyrotron, operating at the second harmonic of the electron cyclotron frequency, are reported. This gyrotron has been designed in order to evaluate new emitter technologies for future high-power fusion gyrotrons. In order to have an opportunity to perform evaluation of such novel emitter technologies several requirements have to be fulfilled in the design of the gyrotron. First, the current density at the emitter surface has to be similar to that in MW class fusion gyrotrons. Second, the gyrotron has to operate in a continuous wave (CW) regime as is the case for fusion gyrotrons. Third, the gyrotron has to operate at a wide range of electron beam parameters in order to have the opportunity to check the emitter performance in different regimes. And finally, fourth, the gyrotron has to be designed in order to have the capabilities to be used in material processing applications at the Institute for Pulsed Power and Microwave Technology (IHM).

As a first step, the design of the cavity is presented. In order to provide stable single mode operation of the gyrotron, simulations using a non-stationary multi-mode code were performed. Optimization of the cavity profile made it possible to suppress spurious modes and to provide stable operation of the main mode at the second harmonic of the electron cyclotron frequency. Simulations of the triode-type electron gun were performed in order to provide good quality electron beams in a wide range of operating parameters. A triode-type magnetron injection gun has been used, in order to increase the range of operating parameters. The profile of the gyrotron collector has been optimized in order to decrease the thermal load on the inner surface and to allow CW operation. The cavity and the collector are water cooled to allow CW operation.

The experimental investigation showed stable CW operation of the gyrotron in the designed main mode $TE_{1,2}$. The triode electron gun and the cavity operated stably over a wide range of operating parameters and allowed to achieve high efficiency of the main mode. Comparison of the experimental and theoretical results showed very good agreement.

Experimentally the following result has been achieved in CW operation: 22.5 kW output microwave power at 23.4 kV electron beam voltage and 2.23 A electron beam current with the world record efficiency of 43 % for CW gyrotrons operated at the frequency of 28 GHz at the second harmonic of the cyclotron frequency.

Contents

List of abbreviations

CW	continuous wave
CRP	controlled porosity reservoir
CRT	cathode ray tubes
CCR	Calabazas Creek Research
IHM	Institute for Pulsed Power and Microwave Technology
KIT	Karlsruhe Institute of Technology
MIG	magnetron injection gun
OFHC	oxygen free high conductivity
RMS	root mean square
RF	radio frequency
SEM	scanning electron microscope
TE	transverse electric

List of used symbols

$\vec{B}$	magnetic flux vector with amplitude B
c	speed of light
c_0	speed of light in vacuum
P_{RF}	output RF-power
J_{em}	emitter current density
k	total wavenumber
$k_{\parallel}$	parallel (axial) wavenumber
$k_{\perp}$	perpendicular wavenumber
r_L	Larmor radius
Ω_c	complex propagation constant
ε_r	relative permittivity
μ_r	relative permeability
m_e	electron mass
e	electron rest charge
$v_{\parallel}$	parallel velocity
$v_{\perp}$	perpendicular velocity
W_{kin}	kinetic energy

g	pitch factor
Δg	pitch factor spread
R_{beam}	electron beam radius
I_{beam}	electron beam current
S_{em}	surface area of an emitter
z_{min}, z_{max}	longitudinal dimensions of coils
R_{min}, R_{max}	radial dimensions of coils
N_c	number of turns in a coil
f_{RF}	RF frequency
U_{0_max}	maximum voltage of the main gyrotron power supply
I_{beam_max}	maximum current of the main gyrotron power supply
R_{cav}	cavity radius
G_{mp}	coupling coefficient
J_m	m-th order Bessel function
ϑ_{mp}	p-th root of the derivative of the Bessel function
$f_{c.o.}$	cut-off frequency
$L_{S.d.}$, $L_{S.u.}$	length of the down- and up-taper sections
L_R	length of the cylindrical cavity section
$R_{c.o.}$	radius of the cut-off section

θ_d, θ_u	down-taper and up-taper angles
$t_{ramp-up}$	ramp-up length of a start-up script
ΔR	electron beam misalignment
U_{beam}	beam voltage
I_{beam}	beam current
U_{cath}	cathode voltage (absolute value)
U_{mod}	modulation anode voltage (absolute value)
L_0	distance between cathode nose and the centre of the magnetic field
ΔL	cathode shift
P_{ohm}	ohmic loss
α	tapering angle of the collector
Z_t	distance between transition from taper to straight part of the collector to the centre of the magnetic field
$tan\delta$	loss tangent
n_w	number of the half wavelength inside the window
α_{Br}	Brillouin angle
d_w	window disk thickness

1 Introduction

1.1 Gyrotron components and principle of operation

The gyrotron oscillator is a vacuum electron tube working as an electron cyclotron maser (microwave amplification by stimulated emission of radiation) [Twi58], [Sch59], [Gap59], [FGPY77]. Gyrotrons can produce radiation with output powers ranging from tens of watts [GGG+65] to tens of megawatts [MZZ+10]. The covered frequency spectrum spans from the lower GHz- to the THz-region. The schematic of a high-power gyrotron is shown in Fig. 1.1. A gyrotron consists of the following main components: electron gun, beam tunnel, cavity, quasi-optical system [Fla12], output window and collector. In the electron gun, electrons are extracted from a ring-shaped emitter as cathode and accelerated towards the beam tunnel and interaction cavity by an electric field generated due to the voltage drop between cathode and anode. The electron trajectories follow the magnetic field lines provided by the surrounding magnet system. In the cavity the electron-cyclotron-interaction takes place and the RF-power is generated. In MW class gyrotrons operated in the continuous wave (CW) regime the quasi-optical output system forms the output beam and provides its separation from the electron beam trajectories [Fla12]. The spent electron beam is absorbed in the collector and the microwaves exit the tube through one or more dielectric output windows [Ber11].

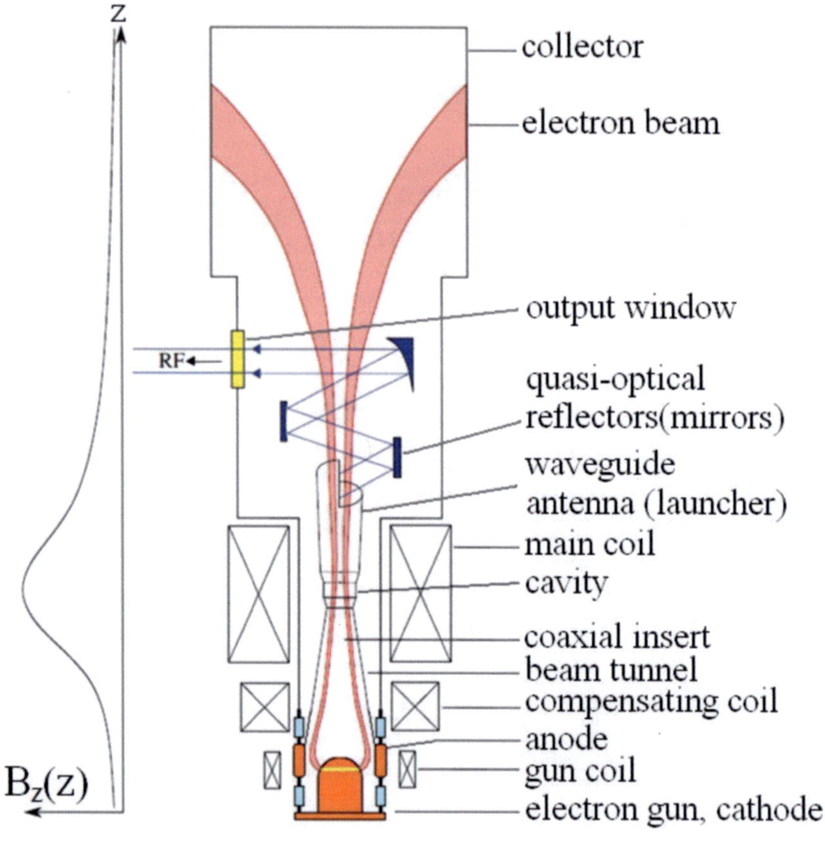

Figure 1.1: Schematic of a MW class gyrotron with quasi-optical output system

The gyrotron configuration shown in Fig. 1.1, is mainly used in high power gyrotrons (P_{RF} > 50 kW in CW), where the collector has a much bigger diameter than the diameter of the magnet bore hole. For low power gyrotrons (P_{RF} < 50kW in CW) another configuration can be used (Fig. 1.2), where the collector diameter is smaller than the magnet bore hole. For this configuration an axial output of the microwave power is used, where the collector serves as the output waveguide.

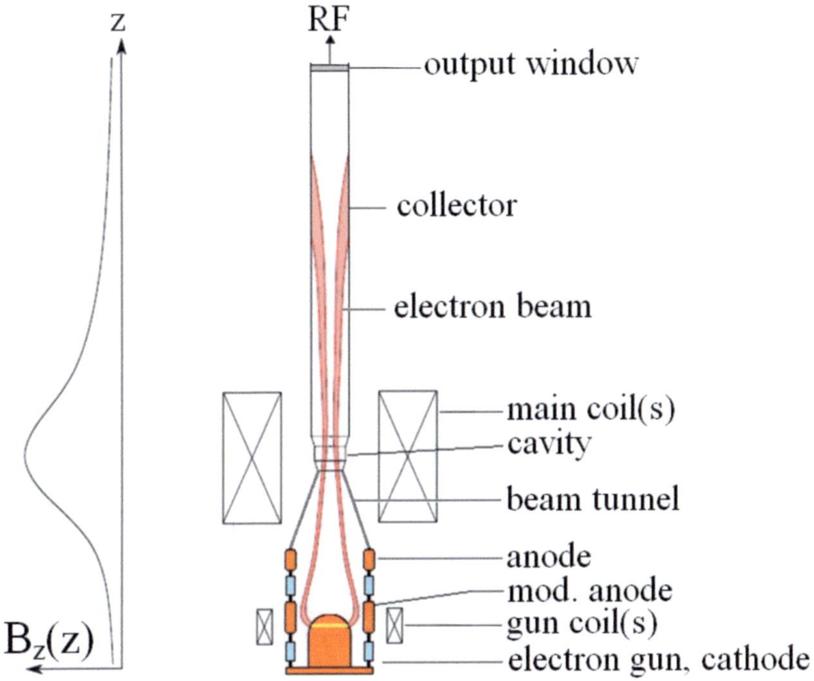

Figure 1.2: Schematic of a 30 kW class gyrotron with axial output waveguide

More detailed information on the gyrotron can be found in [Nus04], [KBT04], [Edg93], [NTP14], [Chu04]. An overview on the state of the art in gyro-devices is given in [Thu15].

In the following sections the main principles and components of the gyrotrons are introduced.

1.2 Electron-cyclotron interaction

Electrons are accelerated from the emitter surface by the electric field provided by the voltage difference between cathode and anode (or modulation anode in case of a triode-type electron gun), and propagate towards the beam tunnel and cavity on helical orbits in the magnetic field $\vec{B}$ produced by the system of magnets. Electrons propagate with the relativistic cyclotron frequency Ω_c and the Larmor radius r_L, which can be calculated by equating the Lorentz- and the centripetal force:

$$\Omega_c = 2\pi f_c = \frac{eB}{m_e\gamma} \tag{1.1}$$

$$r_L = \frac{v_\perp}{\Omega_c} \text{ with } \vec{v} = \overrightarrow{v_\perp} + \overrightarrow{v_\parallel} \tag{1.2}$$

where m_e is the electron mass, e the electron charge, and $v_\parallel$ and $v_\perp$ the parallel and perpendicular velocity components, respectively. Due to the accelerating voltage, the electrons reach a weakly relativistic velocity v with the Lorentz factor γ given by:

$$\gamma = \frac{1}{\sqrt{1-(v/c_0)^2}} = 1 + \frac{W_{kin}}{m_e c_0^2} \tag{1.3}$$

where W_{kin} is the kinetic energy and c_0 the speed of light in vacuum. With an additional RF-field at the angular frequency ω_{RF}, which is nearly synchronous to the rotation of the particles, some electrons lose energy to the RF-field and some gain energy, depending on their relative phase position. While at nonrelativistic energies, the net energy exchange between the entire electron beam and RF-field averages to zero, the changing relativistic factor γ of the individual electrons can cause a non-zero net energy transfer and thus permits the generation and enhancement of the RF-field. The electron-cyclotron-interaction, as described in the following, is therefore a relativistic effect. As a consequence of the changing electron energy and the corresponding cyclotron frequency change, the frequency difference $|\omega_{RF} - \Omega_c|$ between RF-field and electron movement changes and thus the relative phase difference for every particle varies over time. Electrons in an accelerating phase position increase γ, thus decrease their Ω_c and increase the difference with respect to the initial higher ω_{RF}. Over time, they accumulate an increasing phase lead with which they tend to leave this undesired phase position more quickly. In contrast, decelerated particles increase their Ω_c and decrease the difference with respect to ω_{RF}, which means they tend to remain longer in their useful initial phase position relative to the RF-field. In summary, the particles accumulate in a particular relative phase position where they transfer energy to the RF-field. This focusing towards one phase is called "bunching". In order to enable a useful energy exchange, the oscillation frequency ω_{RF} has to be slightly higher than the angular frequency Ω_c of the electrons.

$$\omega_{RF} \gtrsim \Omega_c \tag{1.4}$$

Due to the axial motion $v_\parallel$ of the electrons relative to the RF-field, the frequency ω_{RF} is shifted by a Doppler term. To describe the Doppler term the wave number k is introduced, which gives the number of wavelengths per unit distance.

$$k_0 = \frac{2\pi}{\lambda} \tag{1.5}$$

The total wave number k_0 can be split into two portions for the wave propagation in the axial $k_\parallel$ and perpendicular $k_\perp$ directions.

$$k_0^2 = k_\parallel^2 + k_\perp^2 \tag{1.6}$$

The phase of the field ($\omega_{RF}t$ - $k_\parallel z$) at the position of the electron, which is also moving in the positive z-direction, should be synchronous with the rotational particle movement $\Omega_c t$. As a consequence, the corresponding phase difference vary very little over time

$$\frac{d}{dt}(\omega_{RF}t - k_\parallel v_\parallel t - \Omega_c t) \approx 0 \tag{1.7}$$

Considering the Doppler shift, the resonance condition is derived as:

$$\omega_{RF} - k_{\parallel} v_{\parallel} \approx \Omega_c \tag{1.8}$$

Oscillations are also possible at higher cyclotron harmonics $n\Omega_c$ in which the RF-field is inhomogeneous along the electron rotation. As a consequence, the general oscillation condition is:

$$\omega_{RF} - k_{\parallel} v_{\parallel} \approx n \cdot \Omega_c \tag{1.9}$$

1.3 Electron gun and beam tunnel

The RF-field in the cavity extracts energy out of the perpendicular component of the electron motion. Thus electrons with a high perpendicular velocity component $v_\perp$ are necessary for efficient electron-cyclotron-interaction. In addition, their longitudinal velocity $v_\parallel$ has to be sufficiently high to enable the particles to reach the interaction region and leave it again after energy transfer. In gyrotrons magnetron injection guns (MIGs) are usually used to provide the necessary hollow electron beam [Edg93]. Figure 1.3 shows the cross-section of a typical triode-type MIG.

The electrons are accelerated starting from a heated emitter ring towards the beam tunnel and cavity. According to the characteristics of the magnetic field, the emitter forms a hollow beam. The MIGs used in gyrotrons are usually operated under temperature-limited (TL) condition of the emitter [Edg93], [Pio93]. A non-zero electric field exists at the cathode surface and accelerates all emitted electrons towards the anode (or modulation anode for the triode gun). The beam current depends sensitively on the conditions at the emitting surface and can be adjusted by the temperature of the emitter.

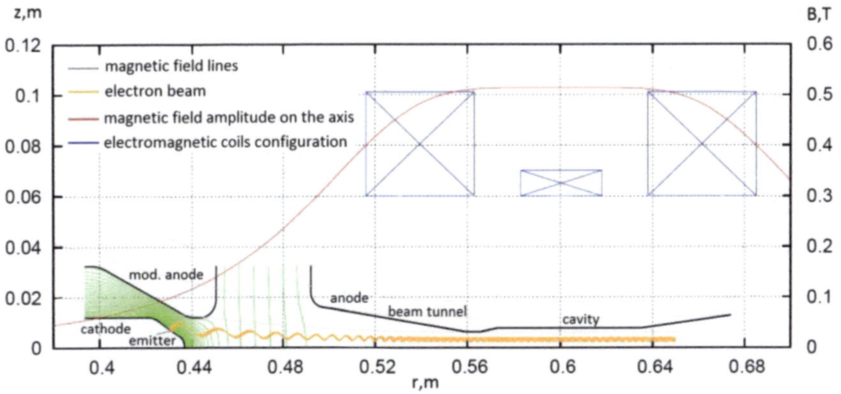

Figure 1.3: Axially symmetric cross-section of a triode-type magnetron injection gun

One of the main characteristics of the electron beam is the ratio between the electrons' perpendicular (rotational) $v_\perp$ and axial (parallel) $v_\parallel$ velocity components. This ratio is also called "pitch factor".

$$g = \frac{v_\perp}{v_\parallel} \tag{1.10}$$

Due to the increase of the magnetic field amplitude from the emitter towards the interaction cavity (Fig. 1.3), the electron beam radius R_{beam} is compressed and energy is transferred towards the perpendicular velocity component, which means that g is increased according to the increasing cyclotron frequency [Edg99], [Nus04]. To describe the pitch factor spread Δg, here the root mean square (RMS) is used [Ber11], [Sch15].

Another important parameter of the electron gun is the emitter current density J_{em} which can be introduced by the following formula:

$$J_{em} = \frac{I_{beam}}{S_{em}} \tag{1.11}$$

where I_{beam} is the electron beam current, and S_{em} the surface area of the emitter.

1.4 Cavity

In the cavity the energy exchange between the helical electron beam and the RF-field takes place due to the electron-cyclotron-interaction. The cavity in gyrotrons is usually a three-section structure as shown in Fig. 1.4: an input down-taper and a uniform middle section followed by an output up-taper. The transitions between each section are smoothed using parabolic contours.

The longitudinal structure of the field inside a gyrotron cavity at its resonance frequency can be described by a complex profile *f(z)* [Ker96]. In Fig. 1.5 one can see the longitudinal field *f(z)* of an operating mode along the cavity with the maximum at the cylindrical cavity section. The transverse structure of the field is given by the eigenmode of the waveguide oscillation in the resonator [Bor91], [JPT+06]. Figure 1.5 shows an example of the TE-eigenmode longitudinal profile and Fig. 1.6 the distribution of transverse electric field in the cavity.

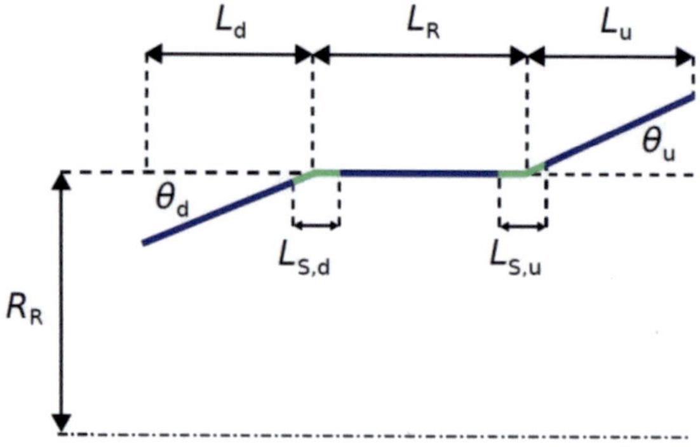

Figure 1.4: Typical gyrotron cavity profile

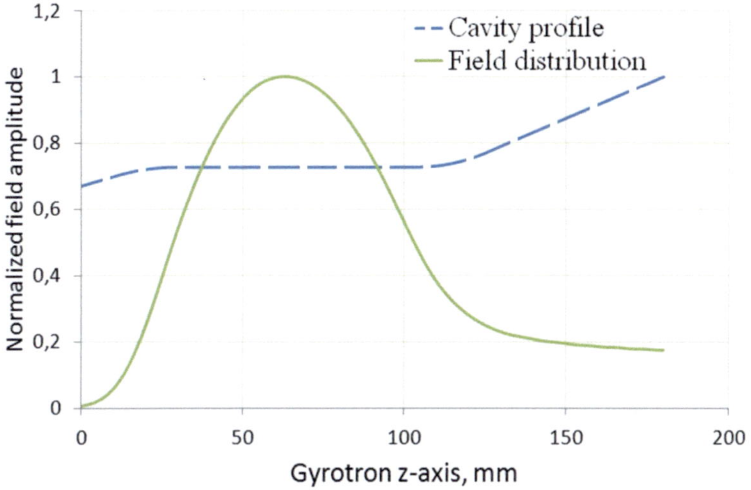

Figure 1.5: Example of geometrical contour and longitudinal field profile of an empty gyrotron cavity

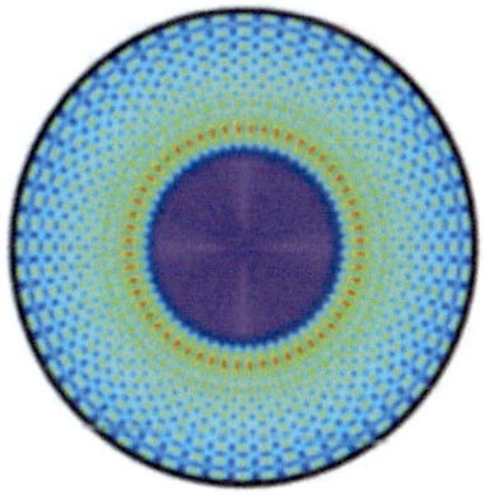

Figure 1.6: Distribution of the transverse electric field (for the $TE_{28,8}$ mode)

1.5 Cathodes

1.5.1 Evolution of modern dispenser cathodes

The family of modern tungsten dispenser cathodes started with the oxide cathode, where $(BaSr)CO_3$ was sprayed onto a nickel base. More details about the oxide cathode history and characteristics can be found in [HW51]. The first step in the development of modern dispenser cathodes was the transition from an oxide coating on nickel to $(BaSr)CO_3$ in a reservoir behind porous tungsten. The cathode was named the 'L' cathode and was developed be Philips Research Laboratories in The Netherlands in 1950 [LJL50]. The L cathode had the following advantages: good emission-current capabilities and a long lifetime. However, it had the following disadvantages: difficulties to provide barium vapour leak-tight seal between the porous tungsten emitter and the molybdenum support body, difficulties to obtain uniform cathode temperatures because of weak heat transfer through the cavity containing the emissive mix to the porous tungsten plug, and a narrow temperature range for good emission.

In 1951 R.C. Hughes, P.P. Coppola, and E.S. Rittner [HCR51] found that a pressed cathode made from barium aluminate and tungsten powder had better performance than the L cathode. The barium aluminate was approximately 70 % BaO and 30 % Al_2O_3, prepared by mixing barium carbonate with aluminium oxide and heating the mixture to about 2000 °C to form the aluminate and free the mixture of the carbonate. This arrangement solved the problems associated with critical processing and activation temperature.

In 1952 it was found that further improvements could be made if the cathode was first formed from porous tungsten and then impregnated with barium aluminate [LH52]. This type of cathode could be easily machined to close mechanical tolerances, had better dimensional stability than pressed cathodes, and the emitter could be made much smoother than pressed emitters. This cathode became known as the Philips 'A' cathode [Lev53].

The Philips B-type cathode developed in 1955, which had additions of calcium oxide to the impregnant that reduced the barium sublimation rate and greatly enhanced the emission properties [Lev55]. The impregnant is $5BaO$: $3CaO$: $2Al_2O_3$ in molar ratio and is sometimes referred to as simply 5-3-2. The 5-3-2 is the most common mix but other variations such as 3-1-1 or 4-1-1 are in widespread use (i.e. $3BaO$: CaO: Al_2O_3 and $4BaO$: CaO: Al_2O_3 respectively).

In [ZV66] it was discovered that over-coating the B-type cathode with a thin layer (200 - 500 nm) of osmium, iridium, ruthenium or rhenium greatly enhanced the emission properties. The effect was verified in [Hou67]. The emission level from a cathode depends on the work function of the emitting surface: the lower the work function, the higher the emission level. The work function is the minimum thermodynamic work (i.e. energy) needed to remove an electron from the surface of a metal. The over-coating greatly reduces the work function from 2.1 to 1.8 eV which results in either a reduction in operating temperature of up to 100 °C for the same emission current density as an uncoated cathode or an increase in emission of well over 300 % for the same operating temperature. It was discovered that a coating of osmium definitely increased emission levels but also had certain disadvantages. For instance, although the cathode lifetime was reasonable, it was less than 100.000 hours as predicted [VK71]. This cathode was named M-type [Sli65], [Bon69], [BBL+69].

The barium scandate cathode in various forms (pressed, impregnated etc.) has been studied since 1967 [FSJ67]. It is essentially a porous tungsten matrix containing barium scandate ($3BaO$: $2Sc_2O_3$) or barium calcium aluminium scandate [VVBZ77].

Some of the best results obtained from cathode research were based on the 'mixed-metal' matrix dispenser cathode (MM cathode). These cathodes have a porous metal matrix made of tungsten and either iridium or osmium and impregnated with barium calcium aluminate. In [Tuc78], it was discovered that the same benefits which are obtained from a sputtered film of osmium or iridium on the surface of the cathode could be obtained by incorporating the enhancing material into the matrix. This improvement made it possible to avoid removal of the osmium or iridium by arcing or ion bombardment. Experiments with cathodes, consisting of a matrix of pure iridium containing barium oxide in the pores, has been demonstrated in [Fal77].

In summary, modern dispenser cathodes currently used in practical vacuum devices include the B, M, and scandate cathodes. The mixed-metal-matrix cathodes are in the development stage [RDKD06], [RKB11]. Research on scandate cathodes has renewed interest in developing cathodes capable of high-current-density emission with reasonable lifetimes [ZZLL04], [LZW+04], [JHZ+04], [Shu04], [YGP+05]. Research is in progress to investigate changes to the emission surface to achieve high-current-density operation at reasonable operating temperatures [SCL+06], [GB04], [SYH04], [IFSW05], [RDKB06], [Hig08], [VJK09], [WLG+09].

1.5.2 Problems associated with M-type cathodes

There are two requirements for production of a high quality electron beam. First, the cathode surface must be uniformly heated to sufficient temperature to release electrons from the surface. The necessary temperature is approximately 1000 °C. It is desirable to achieve temperature uniformity to within a few degrees centigrade over the entire emitting surface. This can be a challenging task considering the support structure, heater, and shielding requirements. Modern, 3D thermo-mechanical analysis programs are now available to precisely simulate the characteristics of the heated materials, including conduction, convection, and radiation heat transfer. Consequently, achieving the required temperature uniformity is now within the designers' control.

A more difficult problem is controlling the work function uniformity of the emitting surface. Porous tungsten cylinders, or billets, are produced by compressing tungsten powder and sintering the material in a high temperature furnace, typically over 2000 °C for 1 – 2 hours. The porosity of the material can be somewhat controlled by the initial grain size of the tungsten powder. Typically, grains are an average diameter of 4 – 5 µm with a wide distribution between sub-micron and more than 10 µm and irregularly shaped. Figure 1.7 demonstrates the structure of a standard dispenser cathode. While it is possible to procure precisely sized grains of spherical shape, the cost is prohibitive for most applications. It has also been found that the performance sometimes suffers with controlled grain size. As a result, there is no precise control of the porosity of most cathodes.

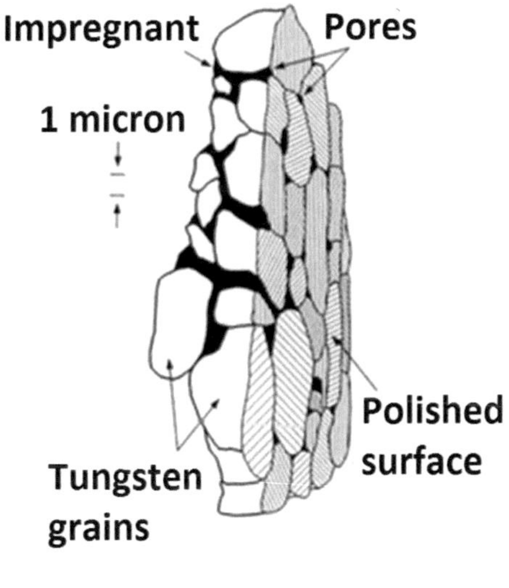

Figure 1.7: Structure of a standard dispenser cathode [JLL00]

Variations in pore size and distribution over the surface can lead to variations in barium diffusion and the subsequent work function. For devices where the emission is limited by space charge at the surface, this variation can be overcome by increasing the cathode temperature. This leads to increased barium diffusion which increases the emission uniformity but decreases the life of the cathode. Once the barium impregnated in the cathode is depleted, the cathode's useful lifetime is ended [GF02].

This solution is not available for cathodes that operate in a regime where the emission is determined by temperature. Cathodes that operate in the temperature limited regime are used in gyrotrons and gyro-amplifiers. Variations in work function generate variations in emission, which can lead to significant performance degradation [AKK+99].

Another problem associated with M-type cathode is their lifetime at high current densities J_{em} (3 – 5 A/cm^2). The usual current density for MW class gyrotrons designed for W7X and ITER is in the range 2.5 - 3 A/cm^2 [JAF+13] and for future DEMO gyrotrons up to 4 A/cm^2 [FPA+15]. Figure 1.8 shows the dependence of the emission current density versus lifetime for B-type and M-type cathodes [Gil86]. As one can see from Fig. 1.8, employing M-type cathodes in space application (where such tubes like klystrons, travelling wave tubes, backward wave oscillators and so on are used) allows to use lower current density J_{em} for cathodes, less than 1 A/cm^2, and provides long life time (more than 15 years). But it can be also seen from Fig. 1.8 that for MW class gyrotrons for fusion applications with usual emitter current densities of J_{em} = 2 – 4 A/cm^2, the lifetime of the cathode is only 1 – 5 years.

One of the most promising cathode technologies that can provide uniform distribution of the electron emission and a long lifetime is a controlled porosity reservoir (CPR) cathode. More details about CPR cathodes are described in the next section.

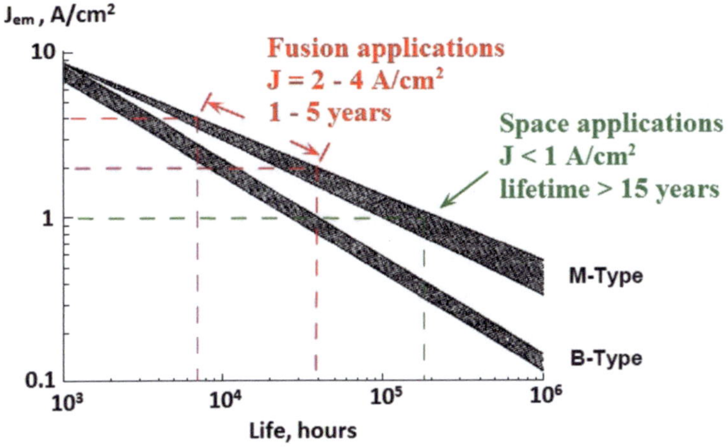

Figure 1.8: Dependence of emission current density from the emitter, for B-type and M-type cathodes [Gil86]

1.5.3 Controlled porosity reservoir cathodes

For more than thirty years, cathode developers have attempted to control the porosity of cathodes, and subsequently the diffusion of barium, in order to control cathode emission. There are numerous patents in this area, though none have achieved success in a practical, production environment [GDGT89], [RCY02], [TG83], [FB86], [GT88], [Wij92].

All approaches involving patterned holes or slots in thin surfaces encounter certain common problems. One problem involves the number and sizes of holes that must be drilled or otherwise created. The "ideal' cathode should have 1 – 2 μm diameter pores about 5 – 10 μm apart. When one calculates the number of holes required for a large cathode, however, the number runs into millions. The time and logistics of creating such a large number of holes over extended surfaces can be a serious problem.

Another problem is assotiated with differential thermal expansion. Thermionic cathodes operate at approximately 1000 °C. Differential thermal expansion between the thin cathode surface and the support structure of the cathode assembly causes buckling and creasing of the surface. This, in turn, leads to non-uniform electric fields on the surface, which results in emission variation.

A number of small, laser drilled CPR cathodes were manufactured for testing in cathode ray tubes (CRTs) about 25 years ago [Fal89], [Fal90], [Fal91], [Fal92]. They initially performed very well. With time, however, the large amount of barium evaporation caused difficulty because of the heavy deposit of barium on various electrodes. The cathodes used very thin laser drilled dispenser caps over a reservoir of barium producing material. The caps were only 25 to 50 μm thick because the laser equipment was incapable of drilling anything thicker through the refractory metals used. The openings were between 20 and 25 μm in diameter on 37 μm centres. There was not a sufficient impedance to restrict or control the amount of barium which essentially blew through the openings into the tubes. Although some devices were lifetime tested for thousands of hours and performed reasonably well despite the high evaporation rate, the effect on certain electrodes made it impractical to continue. The research was redirected to use conventional dispenser cathodes.

The first experiments to produce a metal structure with uniformly distributed pores over the surface and at the same time structure that is thick enough to introduce sufficient impedance to control the amount of barium delivered on the emitter surface has been performed by B.H. Alexander and R.W. Balluffi [AB57]. Their experimental set-up is shown in Fig. 1.9, which consists from a copper spool and a multilayer winding of fine copper wires. Alexander and Balluffi performed experiments that included sintering and evaluating changes in the size and shape of the voids between wires, experiments are described in more details in [Jon60].

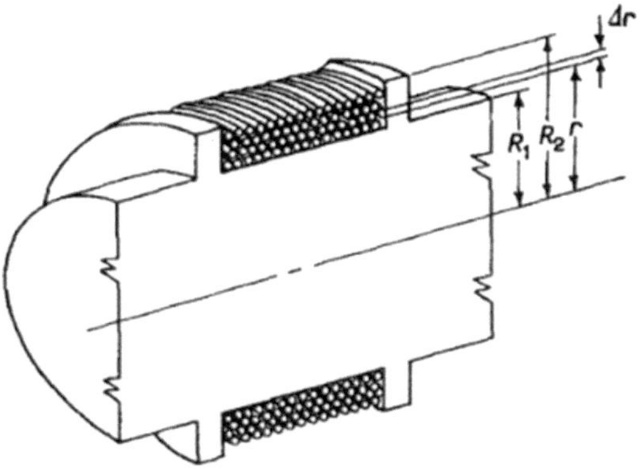

Figure 1.9: Experimental configuration of Alexander and Ballufi [AB57]

Their results indicated that shrinkage is primarily due to volume diffusion of atoms from grain boundaries between adjacent wires to sharply curved surfaces of the voids. The void configuration during various stages of sintering is shown in Fig. 1.10. Section (a) is the initial state prior to sintering where the wires are separate and no diffusion has occurred. Stage (b) is an intermediate stage during sintering. In this stage, atomic diffusion occurs where material moves from the boundary region between wires and from the convex surface of the wires where there is no contact with adjacent wires. It is the small radius surface at the three corners of the void that attracts diffusing material. The final stage is shown in (c).

According to Alexander and Balluffi, this process continues until the grain boundaries between the wires disappear and the void becomes spherical (cylindrical). At this point, further sintering does not result in appreciable changes to the size of the void. The size of the voids can be controlled by the sintering temperature and time. This has implications for cathode manufacture, since it provides an independent mechanism for controlling the pore size and resulting barium diffusion rate. Alexander and Balluffi provide a thorough mathematical treatment of the relationship between time, temperature, and material properties on void characteristics, including comparison with experimental measurements for sintered copper wires.

Experiments were performed by Calabazas Creek Research (CCR) to extend the work of Alexander and Balluffi to tungsten to verify that structures appropriate for CPR cathodes could be obtained. More detailed descriptions of these experiments can be found in [IFSW05], [LFMC10]. The experiments consisted of sintering and evaluating 20-μm-diameter tungsten wire wound on a "rectangular" molybdenum spool. A photograph of the wire wound spool is shown in Fig. 1.11. The spool produced four rectangular tungsten sections that were 18 x 13 x 3 mm. The wire was wound parallel to the 18-mm side; consequently, the cross section of the wires was 13 x 3 mm with voids that penetrated through the 18-mm length.

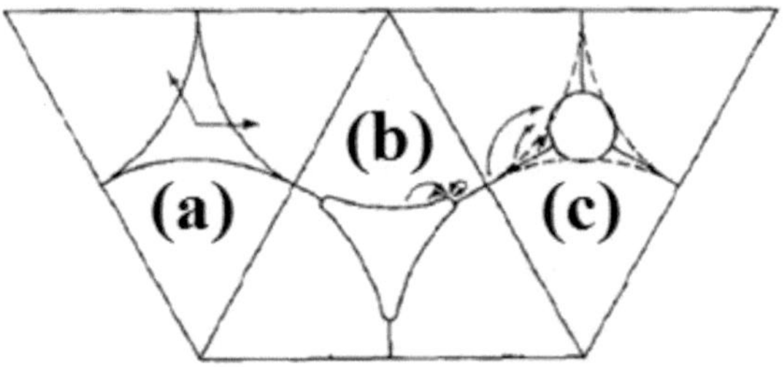

Figure 1.10: Schematic end view of wires showing different stages of sintering. Views are: (a) before sintering, (b) cusped void at early stage, and (c) spherodization [IFSW05]

The structure was sintered at 2075 °C for 75 min. The sections were removed from the spool at the corners and cross sectioned using tungsten electro discharge machining. The face was lapped and examined using a scanning electron microscope (SEM). Figure 1.12 shows sintered tungsten wires formed as a cylinder and Fig. 1.13 is an SEM photo of a region of the cross section. Based on the images in Fig. 1.13, it would appear that this represents an intermediate stage in the sintering process before complete sphericity of the voids is obtained. The voids are approximately 4 μm in maximum dimension with distance between voids consistent with 20 μm diameter wire. The horizontal streaks are from the lapping prior to examination in the SEM. While not fully sintered, the structure appears to represent a more attractive material for cathodes than porous tungsten.

Figure 1.11: Molybdenum spool [LFMC10]

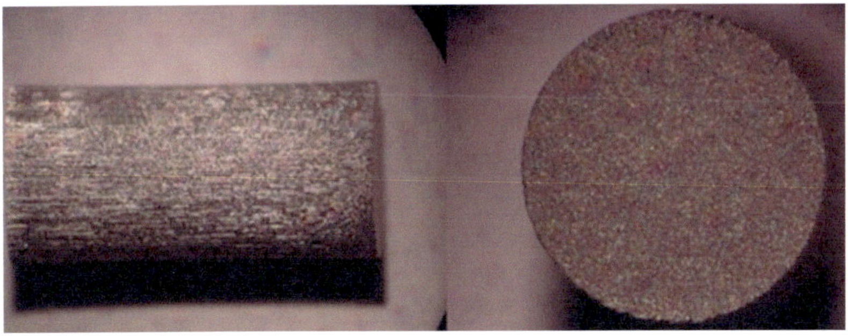

Figure 1.12: Cylinder formed from sintered tungsten wires [LFMC10]

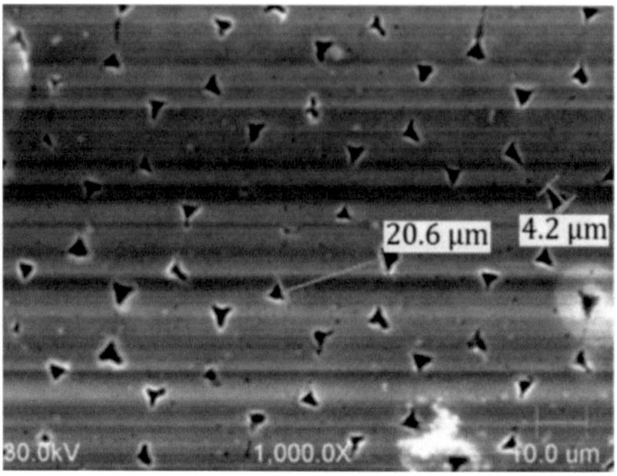

Figure 1.13: SEM photograph of 20 μm diameter tungsten wires sintered at 2075 °C for 75 min. The periodicity is approximately 20 μm with extent of the pores being approximately 4 μm [LFMC10]

The periodicity is approximately 20 μm with extent of the pores being approximately 4 μm. A general configuration of CPR cathode is shown in Fig. 1.14. It consists from heater, a reservoir with barium compound (BaO, CaO, Al_2O_3), and emitter. The emitter is produced from sintered tungsten wires by cutting the proper thickness defined by the configuration of the specific electron gun. A configuration of the CPR cathode which was designed for the gyrotron designed in this dissertation is shown in Appendix A.1.

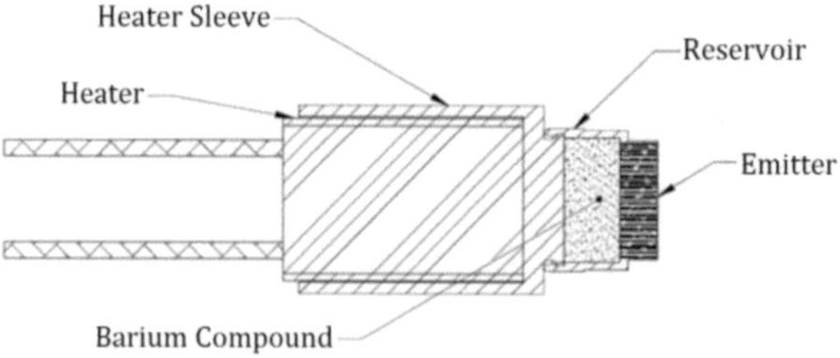

Figure 1.14: CPR cathode configuration [LFMC12]

1.5.4 Advantages of CPR cathodes over M-type cathodes

The most common type of a cathode used in gyrotrons is M-type. In this section, different advantages of the CPR cathodes in comparison to M-type cathodes are presented.

1) **Lifetime performance.** As it was shown in section 1.5.3, construction of a CPR cathode allows to use a big volume reservoir for the barium compound, which can provide enough barium for cathode lifetime of 100.000 hours with emitter current density $J_{em} \approx 3$ A/cm^2 compared to M-type emitter with a lifetime of 10.000 hours with the same emitter current density.

2) **Lifetime prediction**. It was described in Section 1.5.2 that for M-type cathodes the barium compound is impregnated into the voids of the porous tungsten structure of the cathode and the amount of barium strongly depends on the grain size, pore structure and sintering procedure and it is very difficult to predict lifetime taking into account all these factors. In contrast to M-type cathodes, CPR cathodes contain a barium reservoir, where the amount of barium can be easily calculated from the volume of the reservoir and also that barium is easily available from the reservoir through the straight channels of the emitter structure (sintered tungsten wires).

3) **Uniform distribution of barium at the surface of the emitter**. Due to the structure of a CPR emitter, which consists of sintered tungsten wires, distribution of pores between wires is uniform which provides a uniform distribution of barium at the emitter's surface. In comparison, M-type emitters have a non-uniform distribution of pores due to the different sizes of tungsten grains and voids, which gives a non-uniform barium layer on the surface of the emitter.

4) **Constant and controllable diffusion of barium to the emitter surface.** The channels in a CPR emitter structure are formed by sintered tungsten wires and are uniformly distributed, which provides constant diffusion

of barium to the emitter surface. Also by varying the diameter of the tungsten wires and the sintering procedure (temperature and time of sintering), the diameter of the channels through which barium diffuses to the emitter surface can be changed and this allows to control the barium diffusion rate. If one compares it with the construction of M-type emitters it can be noticed that the barium diffusion rate depends strongly on the size and the structure of the channels in the porous tungsten, which depends on the size of the tungsten grains, and it is very difficult to control these parameters, which makes it difficult to control the barium diffusion rate.

1.6 Main goals of the project

The main goals of this work are listed in the following:

1) Development of a CW gyrotron with emitter current density in a range from 0.5 A/cm^2 to 2.5 A/cm^2.
2) Design of a magnetron injection gun, a cavity and a collector for CW regime.
3) Design of a CW gyrotron that can be used for technological applications (materials processing).
4) Construction of the complete gyrotron and the corresponding experimental test stand.
5) Experimental investigation of the gyrotron.

In addition to these main goals the following requirements had to be fulfilled during the project:

6) The CW tube design should be modular (main components should be connected by flanges)
7) The design should be compact.

Each of these requirements can be met in different gyrotrons, but combination of these features in one gyrotron was the big challenge of this work, for example, combination of the 6th and 7th requirement. In order to allow CW operation, the gyrotron should be sealed in order to minimize leakage rate and has to be baked-out in an oven with high temperatures. Modularity at the same time needs flange connection, which means that special type of gaskets should be used in order to withstand the bake-out process, and an additional ion-getter pump has to be used in order to compensate higher leakage rate compared to a sealed gyrotron. The requirement #7 tells us to come to a compact gyrotron design in order to minimize manufacturing cost and accelerate the manufacturing of the gyrotron. As a basis of the compact design of this gyrotron, a configuration which is employed for technological gyrotrons was used in this work [DBG+08]. In order to provide long lifetime of a technological type gyrotron, the emitter current density should be around 1 A/cm^2 [DBE+07], [SBC+11]. However in order to evaluate a controlled porosity reservoir emitter (or new emitter technologies in general) operating in regimes similar to MW class fusion gyrotrons, an emitter current density of 2.5 A/cm^2 has to be used [JAF+13]. Of course, such a high current density for the technological type gyrotron creates additional difficulties for electron gun design and optimization.

2 Physical design of the gyrotron cavity

2.1 Introduction

In order to define a cavity profile which provides stable single mode generation, several cavities with different operating modes have been studied. A gyrotron cavity should satisfy different physical and technical requirements. The main technical requirements are related to the magnet system available at IHM for technological gyrotrons that has to be used in the experimental investigation of the tube. Figure 2.1 shows the magnet system that is composed of two magnets, the gun magnet and the main magnet. The gun magnet consists of one coil and the main magnet includes four coils, three coils (coils #1, #2, #3) that provide a flat top region of the magnetic field distribution, and an additional coil (coil #4) for optimization of the magnetic field profile at the cavity [MGDL08]. The parameters of the coils are listed in Table 2.1, where z_{min}, z_{max} are the longitudinal and R_{max}, R_{min} the radial dimensions of the coils, respectively, N_c represents the number of turns. The gun coil is used for fine adjustment of the magnetic field at the electron emitter region and for controlling electron beam parameters such as pitch factor or beam guiding centre radius. The maximum achievable magnetic field of the magnet is 0.58 T (Table 2.2).

The requirement #3 mentioned in Section 1.6 tells that the design of the gyrotron should be suitable for use in materials processing applications. There are several widely used frequencies for microwave material processing: 24.1, 28 and 30 GHz. Taking into account the maximum magnetic field 0.58 T that can be achieved by the normal conducting magnet, the most appropriate frequency is f_{RF} = 28 GHz. In order to generate this frequency, the gyrotron has to be operated at the second harmonic of the electron cyclotron frequency.

Table 2.1: Coil parameters of the gyrotron magnet system

	z_{min} [mm]	z_{max} [mm]	R_{min} [mm]	R_{max} [mm]	N_c
Coil #1	516.0	563.0	60.0	101.0	384
Coil #2	583.0	618.0	60.0	70.0	116
Coil #3	638.0	685.0	60.0	101.0	384
Coil #4	583.0	618.0	85.0	96.0	116
Gun coil	398.0	470.0	69.0	104.0	116

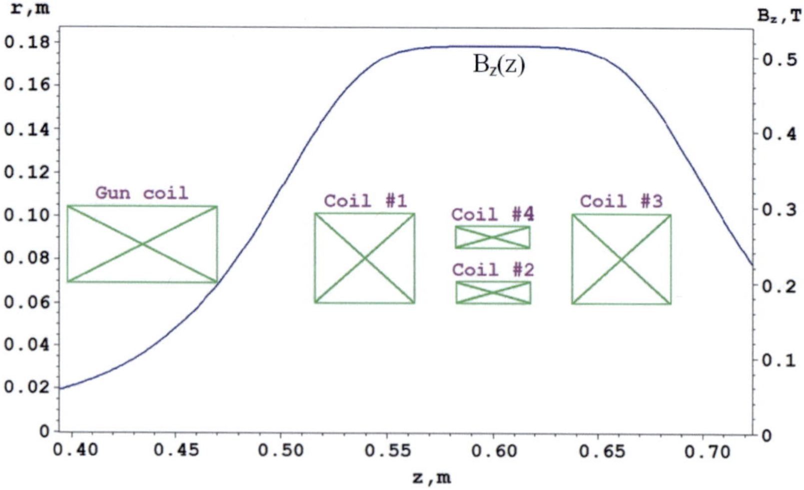

Figure 2.1: Axial magnetic field profile and configuration of the coil system

The main magnet parameters are listed at the Table 2.2.

Table 2.2: Parameters of the normal conducting gyrotron magnet system

Parameter	Value
Maximum magnetic field [T]	0.58
Diameter of the bore hole [mm]	60
Length of the homogeneous field section at (-1 %/ -5 %) level [mm]	90/120

As it was mentioned in Section 1.6, the gyrotron is considered to be an experimental device with a modular structure. This means that the main parts of the gyrotron like electron gun, cavity and collector should be fixed to each other by means of flanges. This increases flexibility during experimental investigations. At the same time a modular design requires flange connections of the gyrotron parts, which limits the maximum cavity radius, because the cavity of course must fit into the bore hole of the magnet. Taking into account a bore hole diameter of D_{bore_hole} = 60 mm (Table 2.2), preliminary design considerations showed that the cavity diameter D_{cav} should not be larger than 32 – 33 mm.

Additional requirements for maximum cathode voltage and electron beam current that can be used in experiments are defined by the main power supply for technological gyrotrons available at IHM (FUG company): U_{0_max} = 26 kV, I_{beam_max} = 2.3 A.

2.2 Cavity selection

As it was mentioned in Section 2.2 the gyrotron has to operate at the second harmonic of the cyclotron frequency. One of the main issues of operation on the second harmonic of the cyclotron frequency is competition with modes at the first harmonic of the cyclotron frequency. Several cavities with different main modes at the second harmonic of the cyclotron frequency have been considered (Table 2.3). In Table 2.3 different cavity parameters are presented: main operating mode, cavity radius R_{cav} necessary for the operating mode at 28 GHz, radius of the guiding centres of the electron beam R_{beam} corresponding to the maximum of the coupling coefficient G_{mp} (2.1) [ZZMS11], and frequency distance Δf between first and second harmonic frequencies (2.2).

The coupling coefficient [ZZMS11] describes the efficiency of the interaction between the mode $TE_{m,p}$ and the electron beam.

$$G_{mp} = \frac{J_{m-n}^2\left(k_\perp \frac{R_{beam}}{R_{cav}}\right)}{J_m^2(\vartheta_{m,p})(\vartheta_{m,p}^2 - m^2)} \tag{2.1}$$

where J_m is a Bessel function of order m, $\vartheta_{m,p}$ the pth root of the derivative of the Bessel function, and $k_\perp = \frac{\vartheta_{m,p}}{R_{cav}}$ is the perpendicular component of the wave number. It should be mentioned that different rotations of the mode relative to the helical motion of the electrons are presented by different signs of the azimuthal index ±|m| ($TE_{+m,p}$ – co-rotating mode and $TE_{-m,p}$ – counter-rotating mode).

The optimum beam radii R_{beam} that are listed in Table 2.3, correspond to the maximum coupling coefficient G_{mp}. An example of the dependence of G_{mp} on beam radius for the $TE_{1.2}$ mode is shown in Fig. 2.2.

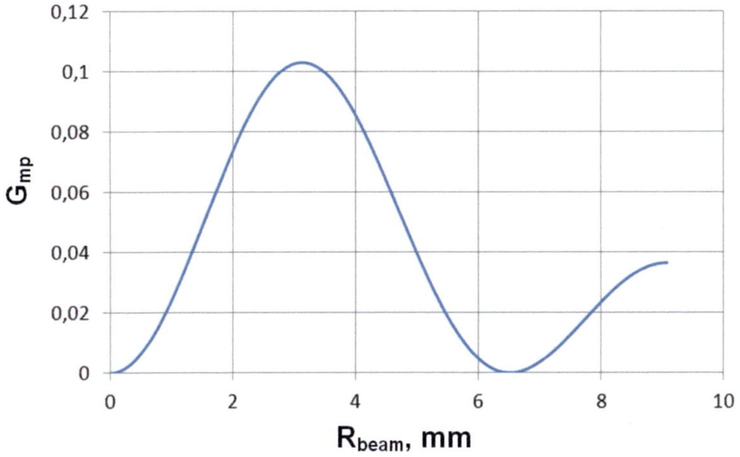

Figure 2.2: Dependence of the coupling coefficient G_{mp} on beam radius for the $TE_{1,2}$ cavity mode

The frequency distance between the first and the second harmonic can be described by the following formula:

$$\Delta f(\%) = \frac{f_{c.o.}(n=2) - f_{c.o.}(n=1)}{f_{c.o.}(n=2)} * 100\% \tag{2.2}$$

where $f_{c.o.}(n{=}1)$ and $f_{c.o.}(n{=}2)$ are the cut-off frequencies (2.3) of the first and the second harmonic modes, respectively.

The cut-off frequency is given by the following formula:

$$f_{c.o.} = \frac{\vartheta_{m,p} \cdot c}{2\pi n R_{cav}} \tag{2.3}$$

Table 2.3: Parameters of the cavities

Cavity main mode	R_{cav} [mm]	R_{beam} [mm]	G_{mp}	Δf [%]
$TE_{3,1}$	7.16	3.13	0.207	12.36 ($TE_{1,1}$)
$TE_{4,1}$	9.06	5.19	0.121	30.75 ($TE_{1,1}$)
$TE_{1,2}$	9.08	3.13	0.103	30.93 ($TE_{1,1}$)
$TE_{0,2}$	11.95	5.23	0.053	12.93 ($TE_{2,1}$)
$TE_{4,2}$	15.8	5.16	0.045	17.36 ($TE_{0,1}$)

The last column in Table 2.3 shows the relative frequency distance to the corresponding main competing mode at the first harmonic of the electron cyclotron frequency, which is shown in parentheses.

Modes with higher order than the $TE_{4,2}$ mode have not been considered because they require a cavity radius > 17 mm, and as it has been mentioned in Section 2.1, cavities with R_{cav} > 17 mm cannot be used.

The spectra shown in Figs. 2.3, 2.4, 2.5, 2.6, and Fig. 2.7 have been used where the dependence of the coupling coefficients G_{mp} from the cut-off frequencies $f_{c.o.}$ are shown for different cavities. The first harmonic mode with the highest coupling coefficient is considered as a main competitor for the main mode in each cavity.

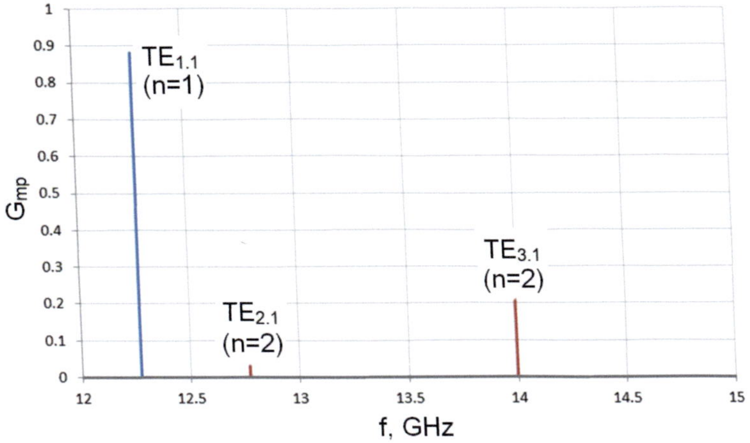

Figure 2.3: Spectrum of the $TE_{3,1}$ mode cavity

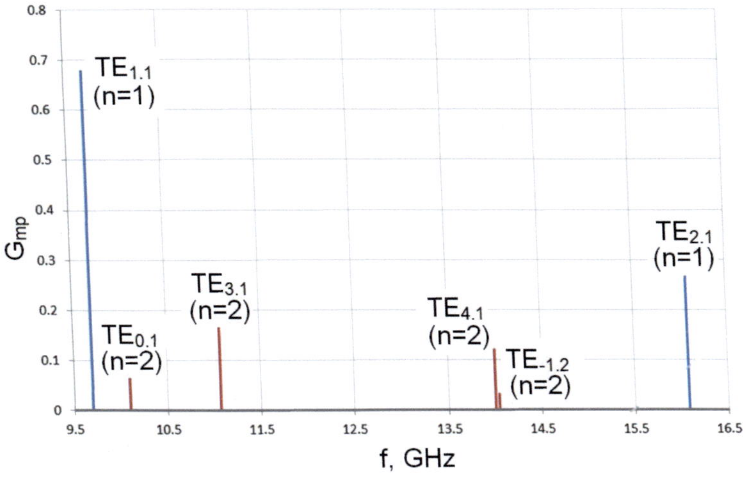

Figure 2.4: Spectrum of the $TE_{4,1}$ mode cavity

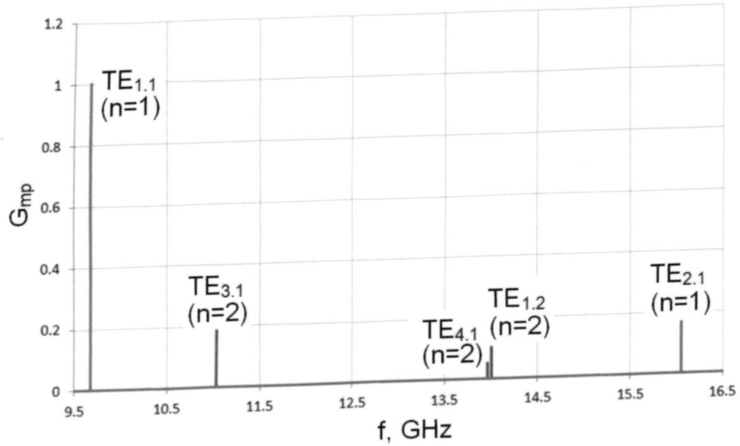

Figure 2.5: Spectrum of the $TE_{1,2}$ mode cavity

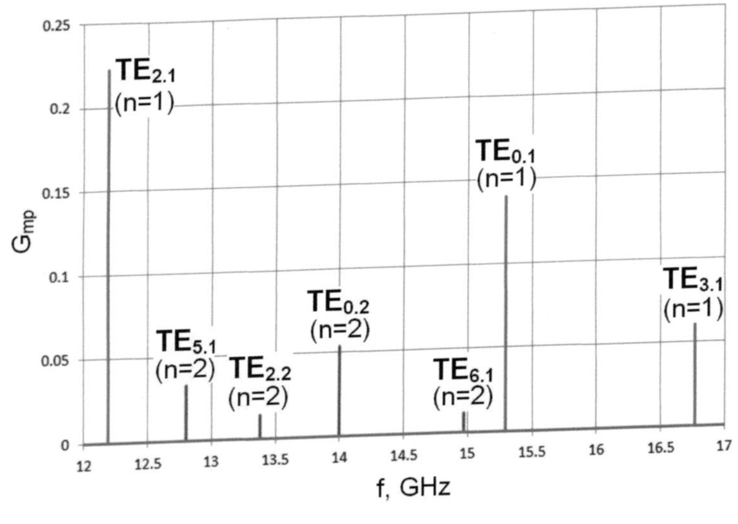

Figure 2.6: Spectrum of the $TE_{0,2}$ mode cavity

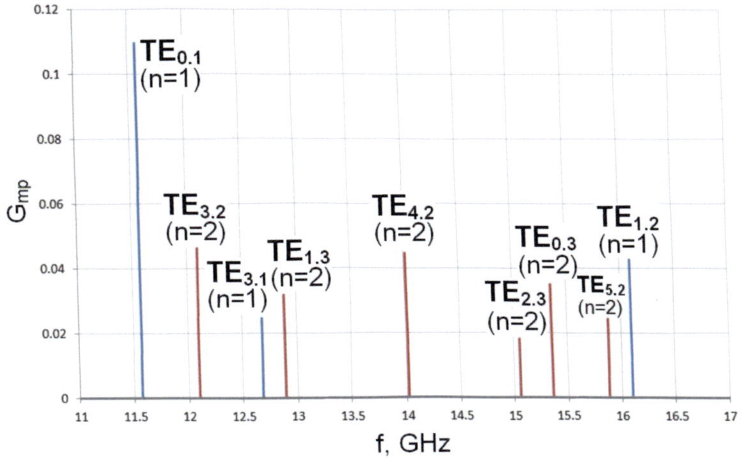

Figure 2.7: Spectrum of the $TE_{4,2}$ mode cavity

As one can see from the results summarized in Table 2.3, the largest distance to the competing mode first harmonic is Δf = 30.93 % for the $TE_{1,2}$ mode cavity. Based on these studies the cavity with the main mode $TE_{1,2}$ has been chosen as the main cavity for the non-stationary multimode simulations.

In Section 2.3 optimization of the cavity profile for the main mode $TE_{1,2}$ is presented. Simulations were done with the non-stationary multi-mode code EURIDICE [APIV12].

2.3 Non-stationary time dependent $TE_{1,2}$ cavity simulations with the EURIDICE code

In the previous section it was shown that the cavity with the main mode $TE_{1,2}$ has the largest distance in frequency from the most dangerous competitor on the first harmonic of the cyclotron frequency. In this chapter the multi-mode non-stationary code EURIDICE [APIV12] is utilized for the optimization of the cavity profile. The main goal of the studies in this section is to optimize the cavity profile in order to provide stable generation of the main mode with the highest possible efficiency.

Parameters of the cavity profile that have been used in the optimization procedure are shown in Fig. 2.8: $L_{S.u.}$ and $L_{S.d.}$ are the lengths of parabolic smoothing towards up or down taper, L_R is the length of the cylindrical cavity section, $R_{c.o}$ the radius of the cut-off section, θ_d the down-taper angle and θ_u is the up-taper angle.

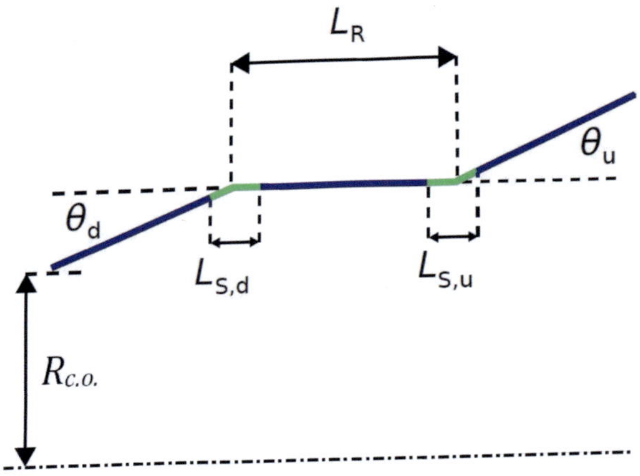

Figure 2.8: Cavity profile (with the parameters that were optimized)

The main goals that have been discussed in Section 1.6 defined that the gyrotron has to be designed in order to be used for two main functions. The first function is the evaluation of new emitter technologies, which requires stable CW operation over a wide range of gyrotron parameters. And the second function is to provide the possibility to use the gyrotron tube for material processing applications. The usual operating parameters for technological type of gyrotrons are the following: frequency 24 to 30 GHz, beam voltage 10 to 20 kV, electron beam current 0.5 to 2 A, output RF power 5 to 15 kW, pitch factor 1.3 to 1.6, and $\Delta g < 10$ %, [SBC+11],

[Thu15]. Based on these two main functions the nominal parameters of the designed gyrotron are presented in Table 2.4.

Table 2.4: Gyrotron parameters

Parameter	Value
Beam current	0.5 – 2.3 A
Cathode voltage	18 – 25 kV
Pitch factor	1.1 – 1.8
Output power	2 – 20 kW
Pitch factor spread	< 10 %

The optimization procedure has been performed by simulating start-up scenarios for different cavity configurations, electron beam parameters and magnetic field profiles, tailored by using the additional coil (coil #4). A start-up scenario has been defined as a linear rise of the cathode voltage. This approach has been used in previous studies [Cho14], [Ber11], [Sch15] and has shown good agreement with experimental studies. The main goal of the optimization procedure was to provide stable, efficient generation of the main mode $TE_{1,2}$ for the whole range of electron beam parameters listed in Table 2.4.

The time length of the start-up scenario ramp up phase has been chosen as $t_{ramp\text{-}up}$ = 1000 ns. Investigations showed that the flat top region where the voltage is constant should be around 1000 ns in order to verify stable operation of the main mode $TE_{1,2}$ or find different instabilities that can be related to competition between modes. In Section 2.4 a more detailed investigation regarding the minimum ramp-up time that can be used in order to have proper simulations of mode excitation is presented.

During the simulations of the start-up scenario the following modes have been taken into account (Fig. 2.5): $TE_{1,2}$ (n=2), $TE_{-1,2}$ (n=2), $TE_{4,1}$ (n=2), $TE_{-4,1}$ (n=2), $TE_{3,1}$ (n=2), $TE_{-3,1}$ (n=2), $TE_{2,1}$ (n=1), $TE_{-2,1}$ (n=1), $TE_{1,1}$ (n=1), $TE_{-1,1}$ (n=1). Even though counter rotating modes have very low coupling coefficient, misalignment of the electron beam can increase the coupling coefficient of different spurious modes and lead to their excitation, as it was shown in [ZZMS11].

An example of the start-up scenario where the first harmonic mode $TE_{1,1}$ is suppressing the main mode $TE_{1,2}$ is shown in Fig. 2.9. The following electron beam and cavity parameters have been used: U_{beam} = 20 kV, I_{beam} = 2.2 A, g = 1.5, Δg = 10 %, L_R = 67 mm, $\theta_d = 3°$, $\theta_u = 3°$, $R_{c.o}$ = 8 mm., $L_{S.d.}$ = 15 mm, $L_{S.u.}$ = 15 mm, R_{cav} = 9.08 mm. One can observe from Fig. 2.9, that after 500 ns both modes, the first harmonic $TE_{1,1}$ and the second harmonic $TE_{1,2}$ mode are excited and increase their power but at the ramp-up time around 750ns the first harmonic mode $TE_{1,1}$ starts to suppress the main mode $TE_{1,2}$ and finally at the end of the voltage ramp-up (t = 1000ns) the competing mode $TE_{1,1}$ at the first harmonic reaches its maximum power and the main mode $TE_{1,2}$ is suppressed to the noise level. The RF power of the first harmonic $TE_{1,1}$ mode is separated in two parts. The first part is the power radiated towards the collector region as is the case for the main mode $TE_{1,2}$, and the second part is the power radiated towards the gun region. As can be seen from the Fig. 2.9 most of the power is radiated towards the gun region, which means that the $TE_{1,1}$ mode is generated in the cavity as a backward wave mode. The low power which is radiated towards the collector is a small reflection of the backward wave from the down-taper region of the cavity.

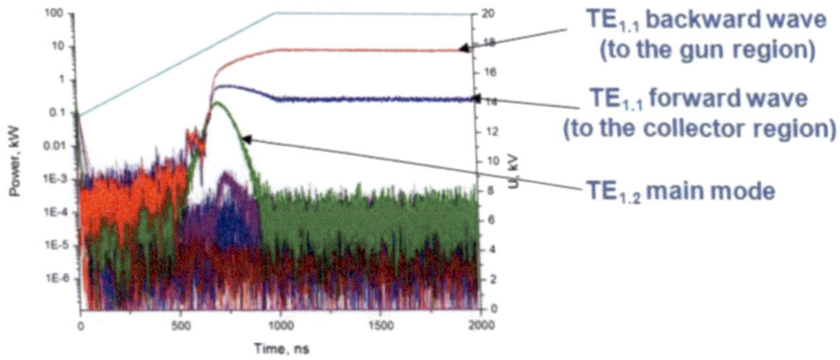

Figure 2.9: Start-up scenario for the $TE_{1,2}$ mode cavity

Fig. 2.10 shows an example of a successful start-up scenario where the main mode $TE_{1,2}$ on the second harmonic of the cyclotron frequency suppresses all spurious modes including the first harmonic mode $TE_{1,1}$ and reaches stable generation at the end of the start-up scenario. The electron beam parameters used were the same as in start-up scenario Fig. 2.9. The following cavity parameters have been used for the simulation of the start-up scenario that is shown in Fig. 2.10: L_R = 117 mm, $\theta_d = 3°$, $\theta_u = 3°$, $R_{c.o}$ = 8 mm., $L_{S.d.}$ = 15 mm, $L_{S.u.}$ = 15 mm, R_{cav} = 9.08 mm. The longer cavity (with L_R = 117 mm) provides more efficient and stable generation of the main mode $TE_{1,2}$ which supresses the competing mode $TE_{1,1}$ at the first harmonic.

An optimization procedure was used to scan through different cavity profiles by using different parameters (Fig. 2.8) in order to provide efficient and stable generation of the main mode $TE_{1,2}$ for the gyrotron parameters listed in Table 2.4. The parameters of the optimized cavity profile are: L_R= 124 mm, $\theta_d = 1.5°$, $\theta_u = 3°$, $R_{c.o.}$ = 8.03 mm, $L_{S.d.}$ = 18 mm, $L_{S.u.}$ = 22 mm.

A detailed drawing of the cavity is presented in Section 6.4.

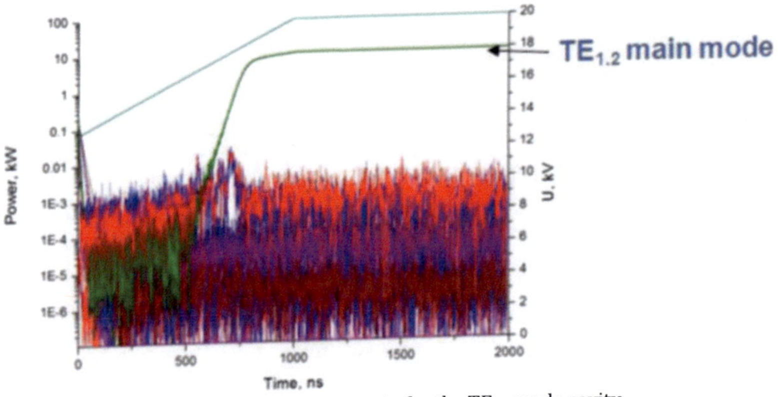

Figure 2.10: Start-up scenario for the $TE_{1,2}$ mode cavity

Simulations of the possible main mode conversion at the up-taper and down-taper region have been performed with the MCONV code from the code package CAVITY [Ker96]. The simulations showed that the mode conversion at the up-taper and down-taper region is less than 0.1 % in power, which is a very low level and should not influence the gyrotron operation.

The maximum efficiency and power regime has been achieved with the following parameters: U_{beam} = 25 kV, I_{beam} = 2.3 A, g = 1.8, $I_{coil\#4}$ = 1.2 A, η = 44.2 %.

2.4 Dependence of the start-up scenario on the rise time

In the previous section multi-mode non-stationary simulations with the EURIDICE code have been discussed. Simulations of the start-up scenarios have been performed with a voltage rise time of t = 1000 ns. In this Section studies of dependence of the mode competition on the rise time during start-up are described. Figure 2.10 shows the start-up scenario with the rise time t = 1000 ns. From Fig. 2.10 it can be seen that the $TE_{1,2}$ mode suppresses all other modes and reaches stable generation at the maximum of the beam voltage. However, if for the same operating parameters and cavity profile the rise time is decreased to t = 500 ns, a very different start-up scenario can be observed. The competing mode $TE_{1,1}$ at the first harmonic of the cyclotron frequency has a higher coupling coefficient (Fig. 2.5) than the $TE_{1,2}$ mode, which provides an advantage during a fast increase of the beam voltage and allows the $TE_{1,1}$ mode to suppress the $TE_{1,2}$ mode. Figure 2.11 shows the start-up scenario with the rise time t = 500 ns. It can be seen that the two modes $TE_{1,2}$ second harmonic and $TE_{1,1}$ first harmonic are excited at the same time at around t = 300 ns. From t = 450 ns, the $TE_{1,1}$ mode starts to suppress the second harmonic $TE_{1,2}$ mode and at the maximum voltage, the first harmonic $TE_{1,1}$ mode reaches stable generation and the main mode $TE_{1,2}$ is suppressed to the noise level.

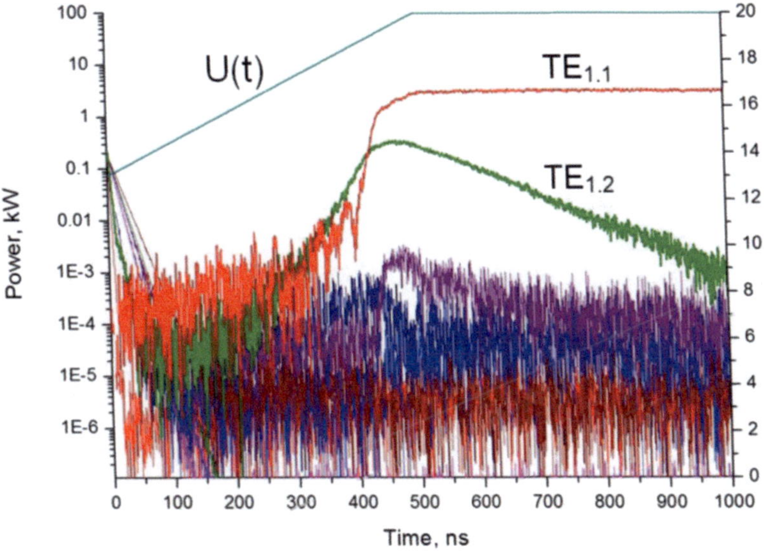

Figure 2.11: Start-up scenario with the rise time t=500ns. The second harmonic $TE_{1,2}$ mode is suppressed by the first harmonic $TE_{1,1}$ mode

These studies showed that not only the magnetic field distribution, cavity configuration and electron beam parameters influence mode competition but also the rise time of the start-up scenario. Simulations with different start-up scenario rise times showed that in order to avoid effects related to the rise time during the start-up scenario, the rise time should not be shorter than 125 ns per 1 kV of voltage rise.

2.5 Influence of electron beam misalignment

In experiments, due to the mechanical tolerances, there is always some degree of misalignment between the electron beam and cavity axes. For the gyrotron in this work misalignment of 0.5 mm has been set up as the maximum value. This means that the generation of the main mode $TE_{1,2}$ has to remain stable if a misalignment of the electron beam with respect to the cavity is less than 0.5 mm. In this section start-up script simulations with misaligned electron beam using the EURIDICE code are presented. For the simulations in this section the electron beam guiding centre spread is set to 0 mm, which defines the electron beam thickness to be two larmor radii ($2r_L$). In Fig. 2.12 the start-up scenario is shown for the same electron beam parameters and cavity profile as it was used in Section 2.3 for Fig. 2.10, with $\Delta R = 0$ mm, where normal excitation of the main mode $TE_{1,2}$ can be observed.

Figure 2.13 shows the start-up scenario with a misalignment of $\Delta R = 1.2$ mm. As it can be seen, strong mode competition between the first harmonic $TE_{1,1}$ mode and the second harmonic main mode $TE_{1,2}$ is present. The analysis of the start-up scenario shows that due to the competition between the $TE_{1,2}$ and $TE_{1,1}$ modes, excitation of the main mode $TE_{1,2}$ starts later (at $t = 1100$ ns) than in the case of zero misalignment ($\Delta R = 0$ mm) of the electron beam (Fig. 2.12), where excitation of the main mode $TE_{1,2}$ starts at around 900 ns. Also it can be noticed that at around $t = 1300$ ns the competing mode $TE_{1,1}$ reaches the same power level as the main $TE_{1,2}$ mode but is then suppressed to the noise level. Also excitation of the counter rotating mode $TE_{-1,2}$ can be observed but with a very small power at around $P_{RF} = 10$ W.

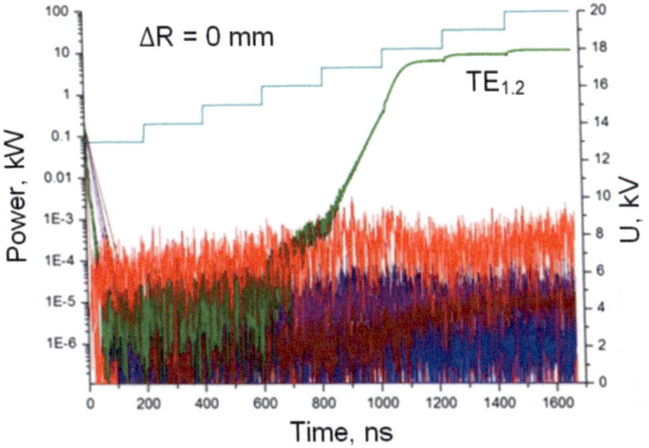

Figure 2.12: Start-up scenario for the $TE_{1,2}$ cavity mode with centred electron beam (ΔR=0)

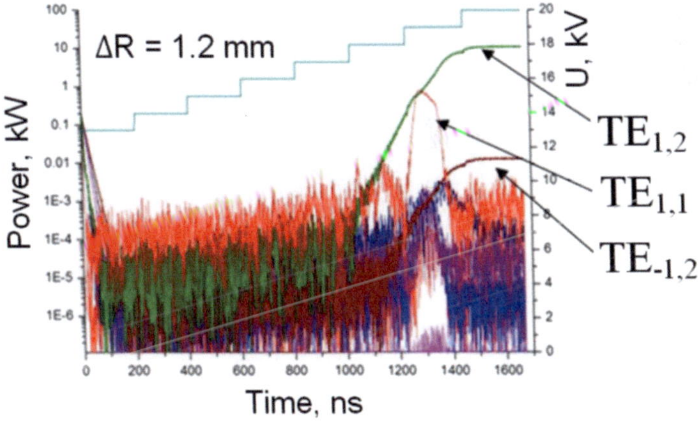

Figure 2.13: Start-up scenario for the $TE_{1,2}$ cavity mode with misaligned electron beam (ΔR = 1.2 mm)

In Fig. 2.14 the start-up scenario for the $TE_{1,2}$ cavity mode with an electron beam misalignment ΔR = 1.4 mm is plotted. One can see that in spite of the fact that the main mode $TE_{1,2}$ is excited earlier (from t = 1100 ns) than the competing mode $TE_{1,1}$, whose excitation starts at t = 1200 ns, the $TE_{1,1}$ mode suppresses the main mode $TE_{1,2}$ and reaches stable operation at the end of the start-up scenario. This investigation showed that in order to get stable excitation and generation of the main mode $TE_{1,2}$ electron beam misalignment, should not be larger than ΔR = 1.2 mm. This fits very well to the technical requirement of ΔR < 0.5 mm and should provide stable generation of the main mode $TE_{1,2}$.

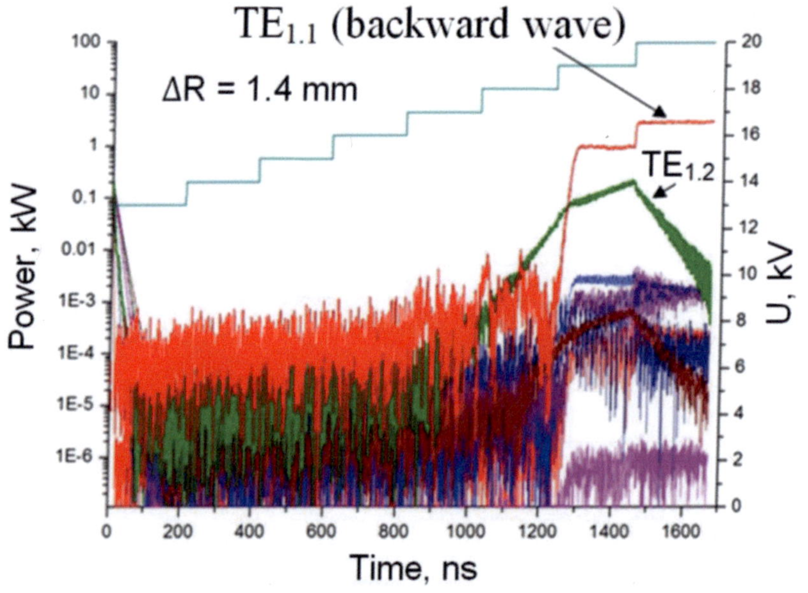

Figure 2.14: Start-up scenario for the $TE_{1,2}$ cavity mode with ΔR = 1.4 mm electron beam misalignment

3 Design of the triode magnetron injection gun

The design of the triode-type magnetron injection electron gun was optimized using the KIT in-house code ESRAY [IB99]. Optimization was done for the cathode, modulation anode and anode profiles in order to minimize the pitch factor spread of the electron beam and to provide an appropriate beam radius in the cavity.

3.1 Optimization of the electrode profiles

In order to perform the optimization, the main design parameters and requirements have to be defined. The requirements can be related to mechanical parameters as well as to the different physical parameters of the electron beam. First of all, the magnetic system available for technological gyrotrons at IHM has to be used. Its layout is shown in Fig. 2.1.

One of the main requirements for the gun design has been mentioned in Section 1.6. The range of the emitter current densities is $0.5\ \mathrm{A/cm^2} < J_{em} < 2.5\ \mathrm{A/cm^2}$. Emitter current densities in a low range $0.5\ \mathrm{A/cm^2} < J_{em} < 1.5\ \mathrm{A/cm^2}$ can be utilized if the gyrotron design is used for technological applications, where the lifetime is a very important factor [SBC+11]. High value of emitter current density $J_{em} \approx 2.5\ \mathrm{A/cm^2}$ could be used for evaluation of new emitter technologies for MW class fusion gyrotrons [JAF+13].

An additional requirement comes from the manufacturing of the CPR emitters that could be tested in this gyrotron in the future. Due to the complicated structure of this CPR emitter (see Appendix A.1), its thickness should not be smaller than 1.7 mm [Ive11]. Because the spread in the guiding centres of the electron beam trajectories at the cavity is mostly defined by the thickness of the emitter, it is preferable to make the emitter as thin as possible to keep the guiding centre spread as low as possible. So the thickness of the emitter has been selected as the minimum that could be achieved in the manufacturing.

Another very important parameter for reliable and stable operation of a magnetron injection gun is the maximum electric field in the electron gun region. Experimental investigations which have been performed at KIT with different gyrotrons showed that in order to avoid arcs and discharges during operation, the maximum electric field in the electron gun region should not be higher than 6 kV/mm [Ber11].

Trapped electrons in the region between cathode and cavity have been studied in [MP09], [KLTZ92] and appear to be one of the main reasons for beam instabilities in the electron gun and beam tunnel and one of the main limiting factors for the increase of the pitch factor as well as a source of increase of pitch factor spread. The task to design an electron gun which has a minimum amount of trapped electrons during operation is very important. As it is shown in [Man06], the number of trapped electrons strongly depends on the type of the electron beam.

As it was shown in previous studies [MP09], [KLTZ92] the alignment of the electron trajectories along the beam path is very important for the performance of the electron gun.

In Fig. 3.1 examples of three different electron beam types are shown [Ber11]. Figure 3.1(a) demonstrates a non-laminar or intersecting flow, where the electrons move along intersecting trajectories, Figure 3.1(b) shows a laminar flow, where the electrons form a thick sheet without any crossing trajectories and Fig. 3.1(c) demonstrates a quasi-laminar or boundary-type flow.

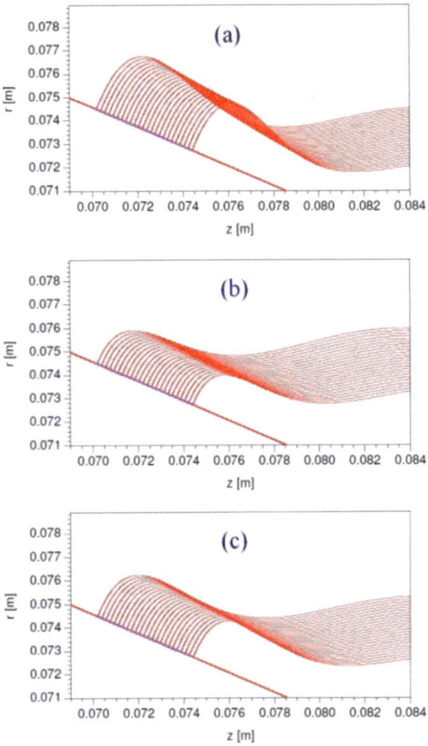

Figure 3.1: Electron beam types: (a) – intersecting type, (b) – laminar type, (c) – boundary type. [Ber11]

This flow is intermediate between laminar and non-laminar flow. All electrons pass through a narrow section (waist) without intersecting each other's trajectory. Operational regimes with regular intersections of the non-laminar electron beam flow can cause an increase in pitch factor spread due to the increase of the space charge forces at the intersections of the electron trajectories. In order to decrease space charge forces during propagation of the electron beam towards the cavity, laminar or quasi-laminar beams are preferable. At the same time a quasi-laminar electron beam has a lower spread of electron beam guiding centres than the laminar electron beam due to the smaller slope angle of the emitter. A smaller guiding centre spread makes it possible to achieve higher interaction efficiency in the cavity and decreases mode competition.

In order to increase the flexibility of the gun operation a triode gun configuration (3 electrodes) has been chosen for the design [Ber11]. Due to the additional electrode (modulation anode) compared to a diode gun, the pitch factor of the electron beam can be changed by changing the modulation anode voltage. In addition, a triode gun is more flexible with respect to cathode position due to the opportunity to adjust the voltage of the modulation anode to compensate a possible axial shift of the cathode. Simulations of the influence of the axial shift of the cathode are discussed in Section 3.2.

Optimization of the gun electrode profiles has been performed with the ESRAY code [IB99]. A special parameter sweeping procedure which will be described below has been used. The main idea of the optimization procedure is to minimize the pitch factor spread by optimizing the geometry of the triode electron gun. In addition to the requirements which have been described above in this chapter, there are several other conditions which have to be fulfilled. First, minimization of the pitch factor spread has to be done for a wide range of pitch factors $1.1 < g < 1.8$ and emitter current densities $0.5\ A/cm^2 < J_{em} < 2.5\ A/cm^2$.

In order to optimize the electron gun, the geometrical profile has to be parametrized. Different dimensions of the cathode, modulation anode and anode have to be optimized to achieve a geometry that provides the lowest possible pitch factor spread for the electron beam parameters listed in Table 3.1.

In Fig. 3.2 the electron gun profile with parameters is shown.

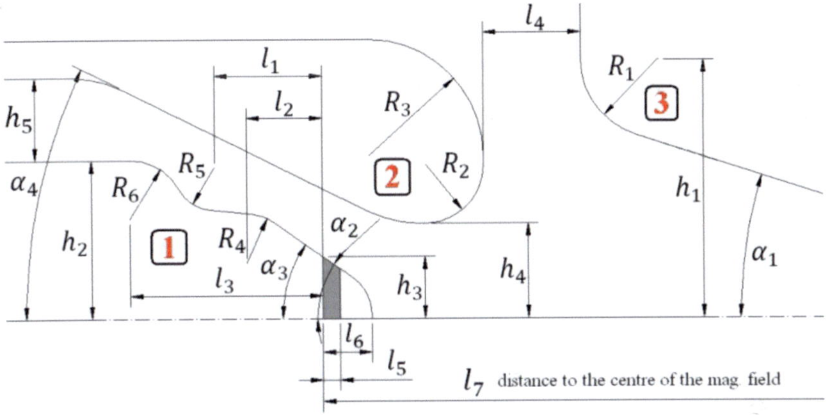

Figure 3.2: Profile of triode-type magnetron injection gun (MIG)

Table 3.1: Cathode and electron beam parameters

Parameter	Value
Beam current [A]	0.5 - 2.3
Cathode voltage [kV]	18 - 25
Beam radius (cavity region) [mm]	3.13
Pitch factor	1.1 - 1.8
Emitter current density [A/cm^2]	0.5 - 2.5
Pitch factor spread [%]	< 10
Emitter width (l_5) [mm]	> 1.7

All parameters and requirements are summarized in the Table 3.1. Due to the fact that the geometry of the gun is described by a large number of parameters (Fig. 3.2), at any simulation step, only 4 different parameters have been taken into account. Scans using four geometrical parameters simplify the analysis of the data that are calculated from one step of the optimization procedure. For each of the four geometrical parameters, the range and number of variations are defined, and each geometry is calculated for different modulation voltages which covers as a minimum the following range of pitch factors $1.1 < g < 1.8$ and also for different electron beam currents $0.5\ A < I_{beam} < 2.3\ A$. Calculation of the average value of pitch factor spread for the pitch factor range is done by using formulas (3.1) and (3.2) for the electron beam current range.

$$\Delta g_{aver.U} = \frac{\sum_{1.1<g<1.8} \Delta g_i}{N_{aver.U}} \tag{3.1}$$

$$\Delta g_{aver.I} = \frac{\sum_{0.5<I_{beam}<2.3} \Delta g_i}{N_{aver.I}} \tag{3.2}$$

$$\Delta g_{aver} = \frac{\Delta g_{aver.U} + \Delta g_{aver.I}}{2} \tag{3.3}$$

where $\Delta g_{aver.U}$ is the average pitch factor spread over the pitch factor range $1.1 < g < 1.8$ (the pitch factor is varied by changing the modulation anode voltage), and $\Delta g_{aver.I}$ is the average pitch factor spread for the electron beam current range $0.5 < I < 2.3$ A.

After calculating the average pitch factor spread from formula (3.3) for each geometry during the scan over four geometrical parameters, the geometry with the smallest Δg_{aver} is chosen and another four geometrical parameters are chosen and the optimization procedure is repeated to define again the geometry with the smallest Δg_{aver}. The optimization procedure is considered to reach saturation when the difference in Δg_{aver}

between current optimization step and previous optimization step is smaller than 0.1 %.

$$\Delta g_{aver}(N-1) - \Delta g_{aver}(N) < 0.1\% \tag{3.4}$$

where N is the step of the optimization procedure.

The following optimized parameters of the gun profile have been found: l_1 = 19.18 mm, l_2 = 9.32 mm, l_3 = 23.65 mm, l_4 = 12 mm, l_5 = 1.7 mm, l_6 = 5.56 mm, l_7 = 169.16 mm, $\alpha_1 = 17.3°$, $\alpha_2 = 34°$, $\alpha_3 = 35°$, $\alpha_4 = 26.0°$, h_1 = 31 mm, h_2 = 19.05 mm, h_3 = 7.7 mm, h_4 = 11.44 mm, h_5 = 9.93 mm, R_1 = 9.5 mm, R_2 = 7 mm, R_3 = 14 mm, R_4 = 6 mm, R_5 = 5 mm, and R_6 = 7 mm.

In Fig. 3.3 dependences of the pitch factor and its spread on $|U_{cath} - U_{mod}|$ for the optimized electron gun profile are shown. U_{cath} is the cathode voltage and U_{mod} is the voltage of the modulation anode. It can be seen that for the required pitch factor range $1.1 < g < 1.8$ a pitch factor spread lower than 10 % has been achieved. Only for pitch factors higher than 2.0 the pitch factor spread is higher than 10 %.

In Fig. 3.4 the dependence of pitch factor spread on electron beam current for the optimized geometry is shown. It can be seen that for the full range of beam currents from I_{beam} = 0.5 A for J_{em} = 0.57 A/cm^2 to I_{beam} = 2.3 A for J_{em} = 2.6 A/cm^2 the requirement to get a pitch factor spread less than 10 % has been achieved.

The layout of the optimized electron gun and cavity is illustrated in Fig. 3.5.

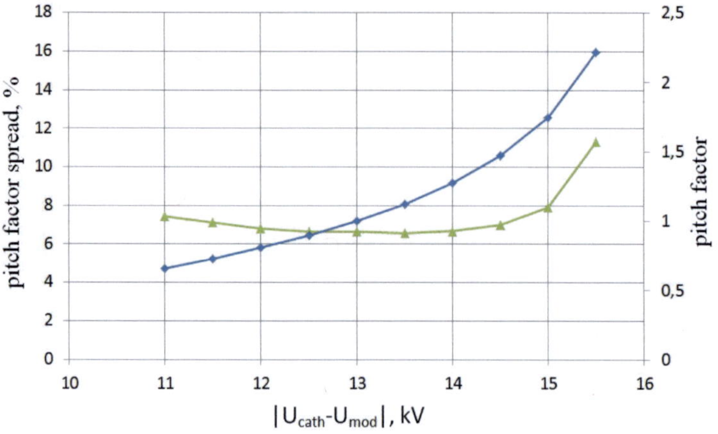

Figure 3.3: Pitch factor spread (triangles) and pitch factor (diamonds) as a function of $|U_{cath} - U_{mod}|$

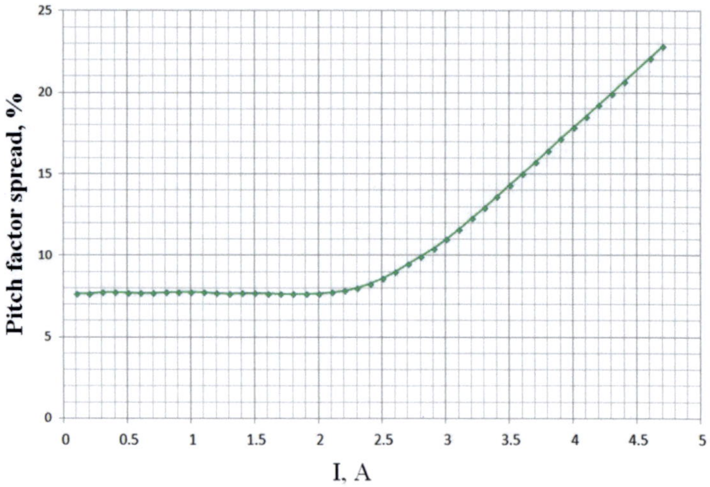

Figure 3.4: Pitch factor spread as a function of beam current for the optimized triode gun profile

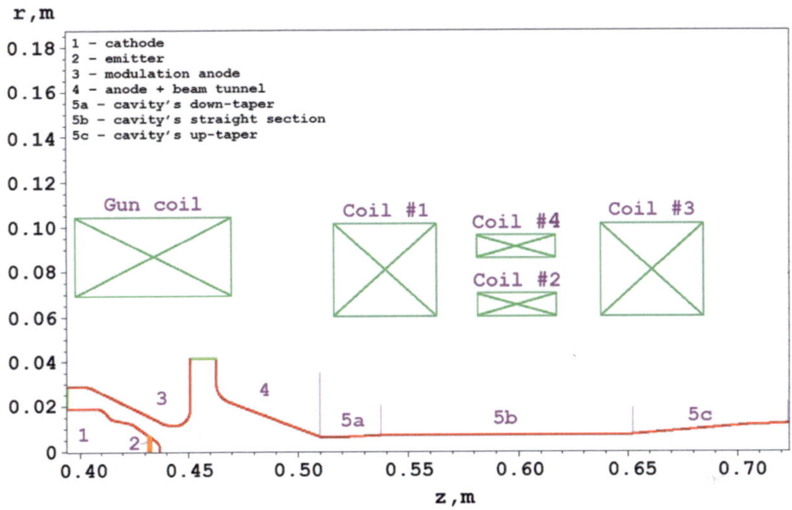

Figure 3.5: Triode electron gun configuration and configuration of the magnet coils

In addition to minimizing the pitch factor spread for a wide range of parameters which has been described above, a very important requirement has been defined in order to avoid trapping of the electrons in the gun region. These electrons can be produced by the metallic surface of the cathode surface. This effect has been described by Pagonakis and Vomvoridis in [PV04]. Electrons that are emitted from the cathode and modulation anode surface (which is also under negative potential) are following the magnetic field lines.

In order to check the possibility of the cathode and modulation anode surface to emit electrons which can be trapped, simulations of the electron gun when all the surfaces of the cathode and modulation anode are considered as possible emitters have been performed with the ESRAY code. Simulations showed that no trapped electrons have been observed in the optimized electron gun.

The applied electric field distribution is shown in Fig. 3.6. As it was mentioned above in order to avoid arcing in the gun region, the maximum electric field should not exceed 6 kV/mm. As one can see from Fig. 3.6, the maximum electric field strength at the gun region is 3.73 kV/mm for a cathode voltage of 20 kV. For the maximum cathode voltage of U_{cath} = 25 kV the maximum electric field at the gun region is 4.67 kV/mm, which is lower than the threshold of 6 kV/mm.

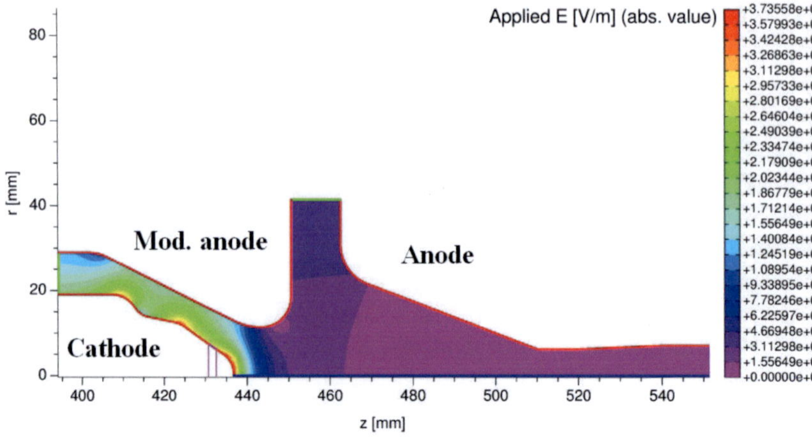

Figure 3.6: Applied electric field (V/m) in the triode gun

The electron trajectories of the optimized electron gun are shown in Fig. 3.7. It can be noticed that at the waist of the electron beam (at the first turn) there are intersections between electron trajectories with a very small angle (between directions of different electron trajectories). The attempt to provide a purely laminar beam (without any intersections of the electron trajectories) during the optimization procedure led to a significantly higher pitch factor spread Δg > 13 % than for the optimized beam with slight intersections. Based on the results of the optimization procedure the decision has been made to keep a low pitch factor spread and an electron beam type with intersection. It will be shown in Section 7 that it was possible to achieve stable operation for a wide range of pitch factors and electron beam currents.

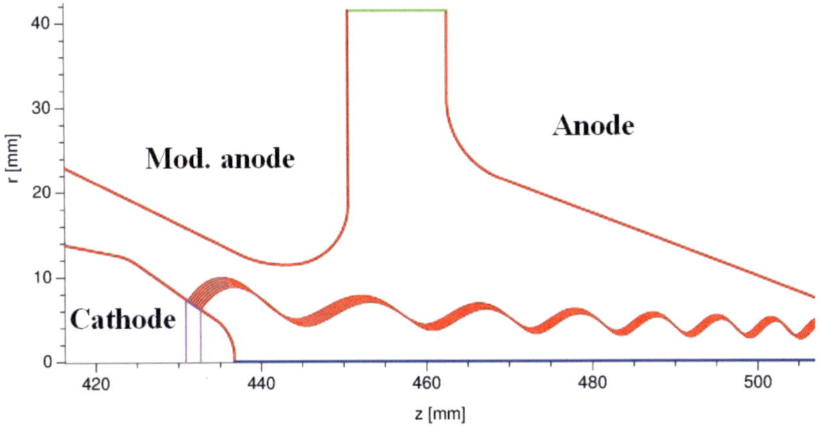

Figure 3.7: Triode type electron gun with electron trajectories

3.2 Influence of imperfect geometry of magnetron injection gun

In order to define the sensitivity of the electron gun to deviations of the different parameters of the geometry, simulations using the ESRAY code have been performed. In experiments, due to the heating of the emitter to temperatures reaching 1150 °C, the cathode can expand up to 0.5 to 1 mm [Ive11]. This means that the emitter position with respect to the modulation anode and anode changes and can affect the electron beam parameters. In Fig. 3.8 the electron gun profile is presented. The distance between cathode nose and centre of the magnetic field is L_0. In the simulations this value has been changed by ΔL. The positions of the modulation anode and anode were fixed for all shifts ΔL.

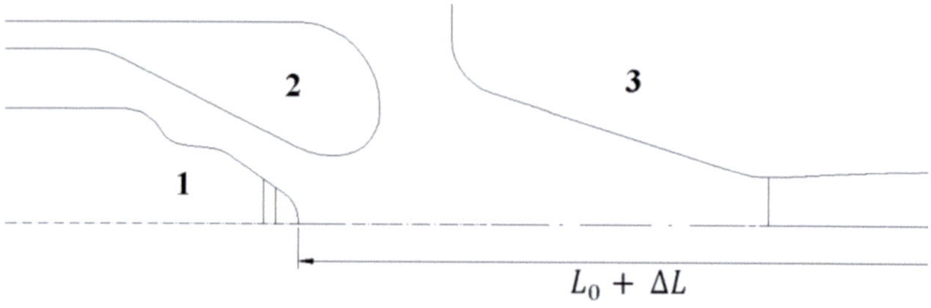

Figure 3.8: Electron gun profile: 1 - cathode, 2 – modulation anode, 3 – anode / beam tunnel. $L_0 + \Delta L$ is the distance between cathode nose and center of the magnetic field

Figure 3.9 shows the pitch factor spread as a function of pitch factor for different cathode shifts from ΔL = -3 mm to ΔL = 2 mm in steps of 1mm. From Fig. 3.9 it can be seen that for shifts 1 mm < ΔL < -1mm the spread is smaller than 10 % for the whole range of pitch factors $1.1 < g < 1.8$ which satisfies the requirements.

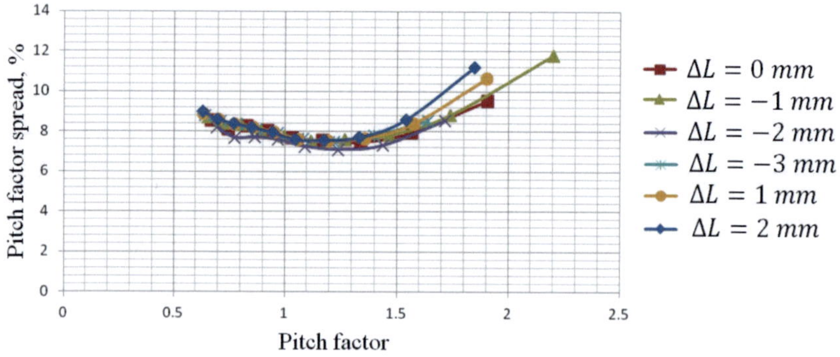

Figure 3.9: Pitch factor spread as a function of pitch factor for different shifts ΔL of the cathode with respect to modulation anode and anode

4 Collector simulations

The optimization of the electron collector included optimization of the angle between the cylindrical part of the collector (constant radius) and the tapered (conical) part (Fig. 4.1) and also the position of the transition point between these two sections. Optimization was performed using the ESRAY code. Simulation of the electron beam with the ESRAY code showed (Fig. 4.1) that using a collector with a constant radius at the part where the electron beam is dissipated leads to a very high wall loading of $P_{ohm} \gtrsim 500$ W/cm², as it can be seen from Fig. 4.1. For reliable operation and long collector lifetime, the wall loading using copper as the absorption material should be lower than 500 W/cm² [Ber11]. In the optimization procedure, two parameters have been used, the angle of the tapering at the beginning of the collector and the position of the transition point between straight part and tapered part of the collector. The area where the electron beam is absorbed with a non-optimized constant collector radius is $0.278 \text{ m} < z < 0.57 \text{ m}$. The main goal of the optimization procedure was to decrease the maximum wall loading by optimizing the profile of the collector. Simulations for different combinations for the tapering angle and the position of the transition point have been performed. The optimized profile which demonstrated the lowest peak wall loading is presented in Fig. 4.1 and the corresponding wall loading in Fig. 4.2 (green colour). As can be seen from Fig. 4.2, tapering of the collector profile at the beginning redistributes the electron beam to the following area $0.263 \text{ m} < z < 0.52 \text{ m}$, which decreases the peak wall loading to around 300 W/cm². Most important is the redistribution of the highest peak to a wider collector area, which decreases the peak wall loading.

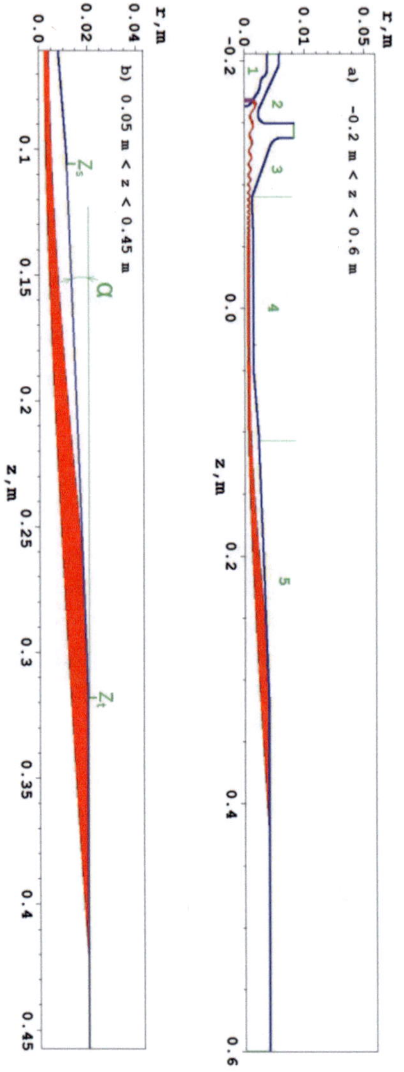

Figure 4.1: Electron trajectories in the collector. 1 – cathode, 2 – modulation anode, 3 – anode-beam tunnel, 4 – cavity, 5 – collector

The optimized collector parameters are: tapering angle α = 2.48° and distance between transition from taper to straight part of the collector to the centre of the magnetic field Z_t = 319.3 mm. A more detailed drawing of the collector is shown in Section 6.5.

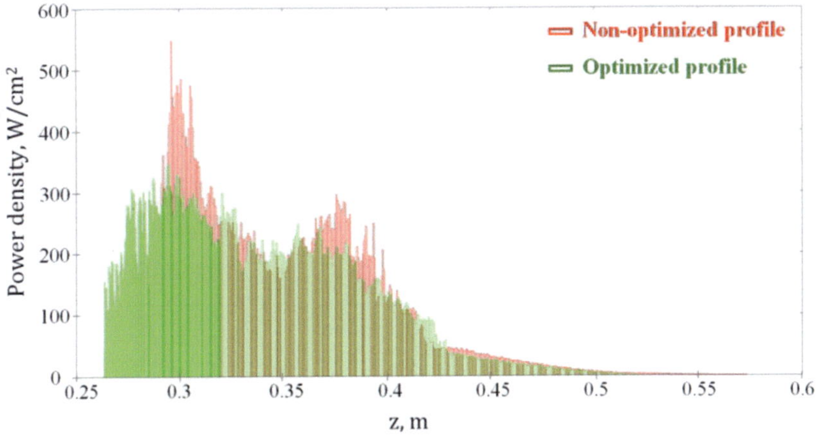

Figure 4.2: Peak power loading along the collector, for the optimized (green) and non-optimized (red) electron-beam deposition profile

A reduction of the peak wall loading from 500 W/cm² to 300 W/cm² will significantly reduce the thermo-mechanical stress on the collector wall. Calculation of the wall loading has been determined considering three generations of secondary electrons which is also necessary to calculate the optimum position of the output window, which should not be hit by primary or secondary electrons.

5 Output window

An output window is used to secure the vacuum inside the gyrotron and to allow RF power to be radiated to the load (e.g. a calorimeter for power measurements). An output window should also be able to survive a bake-out procedure. There are many different dielectric materials that can be used as a window disk. The most widely used materials are: alumina, boron nitride and quartz. Quartz is the most expensive of all three and it is the most challenging to braze inside the waveguide. Boron nitride is a cheap material but it has a much higher loss tangent than alumina (the loss tangent for boron nitride is $5 \cdot 10^{-4}$ and $9 \cdot 10^{-5}$ for AL-995 type alumina). In MW class gyrotrons CVD diamond windows are used due to their high thermal conductivity, but this type of material is very expensive and cannot be used for this work. Taking into account the characteristics of different materials, alumina has been chosen as the material for the output window.

Alumina is unique in having properties that offer a good mechanical-thermal-electrical insulation compromise:

- good mechanical strength under compression load
- good hardness and wear resistance
- good corrosion resistance (liquids and gas)
- good thermal stability
- excellent dielectric properties (in the range from direct currents to microwaves)
- low dielectric constant and loss tangent

Alumina is produced by firing a tightly packed powder form of Al_2O_3 which includes some binder material. Commercially available grades range from 90 % up to 99.95 % with the higher purity generating somewhat higher hardness. In this work AL-995 type of alumina is used which contains 99.5 % of alumina and has one of the best cost - performance characteristics. AL-995 has the following dielectric characteristics at around 28 GHz: relative permittivity $\varepsilon_r = 9.37$, loss tangent $tan\ \delta = 9 \cdot 10^{-5}$ [Ive12].

In order to eliminate reflections from the window its thickness has to be a multiple of a half wavelength inside the window, in this case the wave from the first reflection (vacuum – side) will be cancelled by the wave from the second reflection (air – side) (Fig. 5.1). Figure 5.1 shows a schematic preliminary technical drawing of the window section.

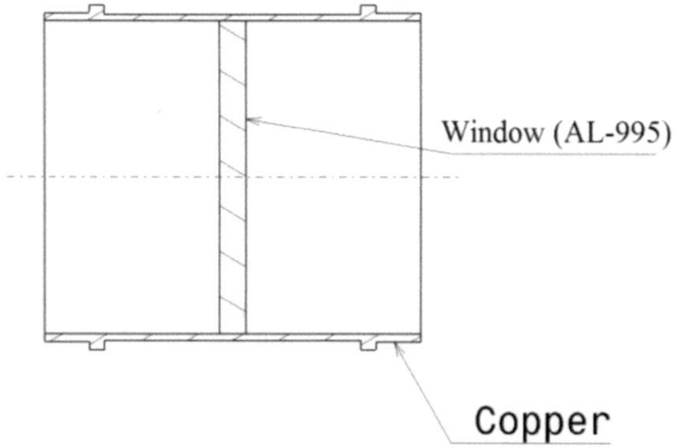

Figure 5.1: Scetch of the output window fixed (brazed) inside a waveguide

It can be seen from Fig.5.1 that the window should be brazed inside a copper waveguide which has the same diameter as the output diameter of the collector. This design allows propagation of the main mode from the collector to the load without reflections.

The formula to calculate the thickness of the window which allows the elimination of reflections is the following:

$$d_w = \frac{n_w \cdot c_0}{2f\sqrt{\varepsilon_r} \cdot cos(\alpha_{Br})} \tag{5.1}$$

where n_w is the number of half wavelengths inside the window, c_0 the speed of light in vacuum, f the RF frequency, α_{Br} the Brillouin angle inside the window of the main $TE_{1,2}$ mode and ε_r the permittivity of the window material.

The Brillouin angle is given by the following expression:

$$\alpha_{Br} = arctan\left(\frac{k_\perp}{k_\parallel}\right) \tag{5.2}$$

where $k_\perp$ and $k_\parallel$ are the perpendicular and the axial wavenumber, respectively, with

$$k_\perp = \frac{\vartheta_{mp}}{R_{cav}} \tag{5.3}$$

$$k_\parallel = \sqrt{k_0^2 - k_\perp^2} \tag{5.4}$$

and

$$k_0 = \frac{2\pi f}{c_0} \tag{5.5}$$

Using (5.1) α_{Br} can be calculated for the $TE_{1,2}$ mode propagating along the waveguide with a radius of 20 mm (collector radius) in vacuum: $\alpha_{Br} = 27.017°$. When the wave moves from one medium to another, it changes the propagation direction, i.e. it is refracted. If it moves from a medium with refractive index n_1 to one with refractive index n_2, with an incidence angle to the normal of the surface normal of θ_1, the refraction angle θ_2 can be calculated from Snell's law.

$$n_1 \sin\theta_1 = n_2 \sin\theta_2 \tag{5.6}$$

The refractive index of a dielectric material is given by

$$n = \sqrt{\varepsilon_r \mu_r} \tag{5.7}$$

where μ_r is the relative permeability (μ_r = 1 for AL-995).

Taking into account the propagation of the $TE_{1,2}$ mode from the vacuum side into the window: n_1 = 1, θ_1 = 27.017°, n_2 = 3.061 using (5.7) one can calculate θ_2 = 8.534°.

In Table 5.1 different window thicknesses with different numbers of half-wavelengths n_w are presented.

From the mechanical point of view an output window with a thickness smaller than 3 mm is too thin to sustain stresses during the bake-out procedure and has a high probability of window cracking. Discussions with CCR led to a decision that the most appropriate thickness of the window is 5.306 mm for n_w = 3.

Table 5.1: Window thickness for different number of half-wavelengths n_w

n_w	d_w [mm]
1	1.769
2	3.537
3	5.306
4	7.074

6 Mechanical design of the gyrotron and the test stand

6.1 Gyrotron design

The mechanical design of the demountable gyrotron has to fulfil several requirements. First, the gyrotron has to run in CW operation. Therefore, the tube has to be baked out in order to achieve clean inner surfaces. In order to have the opportunity to bake-out the tube, all flanges have to be sealed with special Helicoflex type gaskets which allow high temperature bake-out. During operation, the vacuum inside the gyrotron is maintained by two ion-getter pumps. Of course, in order to allow CW operation, water cooling has to be applied to all parts of the gyrotron which will be heated during operation.

The cooling system consists of the following water loops:

1) Cavity cooling, which consists of four inlet pipes with 10mm inner diameter.
2) Collector cooling which is the continuation of the water flow from the cavity cooling loop and in addition to the cavity loop contains another four inlet pipes with 20mm inner diameter.
3) Indirect modulation anode cooling by cooling of the flange to which it is connected.
4) Output window cooling.

Figure 6.1 shows the technical drawing of the gyrotron with the main magnet and the electron gun magnet. The main components of the gyrotron are: triode-type magnetron injection electron gun, beam tunnel, cavity, collector, output window and two ion-getter pumps. The triode electron gun shown in Fig. 6.2 consists of the cathode (Figs. 6.3 – 6.5) with a special feed-through flange for the twelve segmented emitter elements (see Appendix A.1), modulation anode (Fig. 6.6), and anode (Fig. 6.7). The anode includes the beam tunnel (Fig. 6.8), which is done by irregular cuts in a stainless steel cone at the region close to the cut-off section of the cavity. The output cone of the tube after the window is designed in order to provide a smooth transition between the output waveguide with the brazed output window and the calorimeter for measurement of the output power. The second purpose of the output cone is to be a fixture for the rectangular waveguide which takes a small signal from the output wave and directs it to a spectrum analyser in order to get information on the frequency of the output RF wave. In addition, the output cone contains a fixture for an arc detector which has to switch off the main power supply in case of arcing events. This is mainly done to protect the output window against possible cracks initiated by arcs.

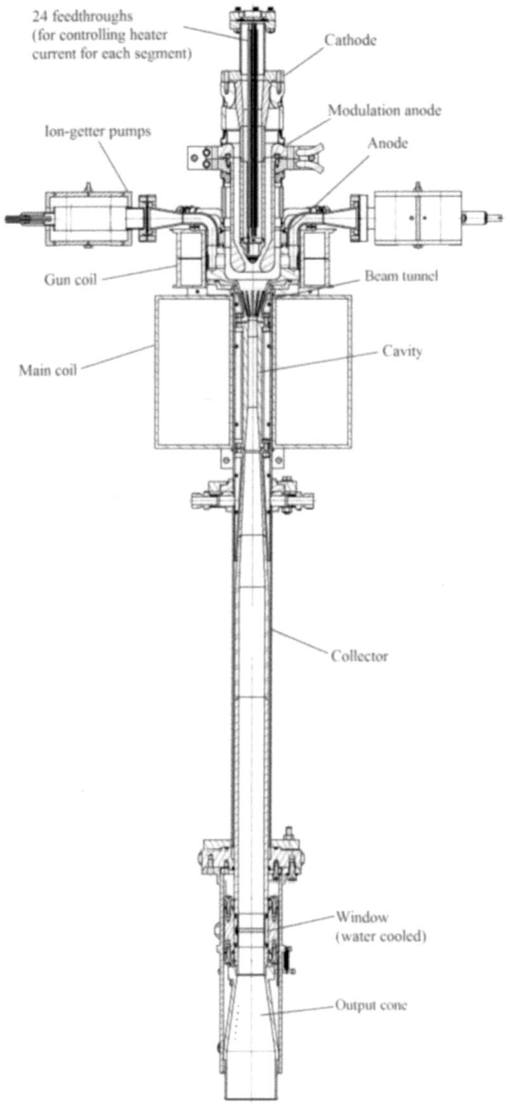

Figure 6.1: Design drawing of the 20 kW/ 28 GHz gyrotron (with main magnet and gun coil)

6.2 Triode type electron gun

In this chapter the different electron gun components will be discussed. Figure 6.2 shows the configuration of the triode magnetron injection electron gun. The preliminary technical drawings of the electron gun have been designed by the author.

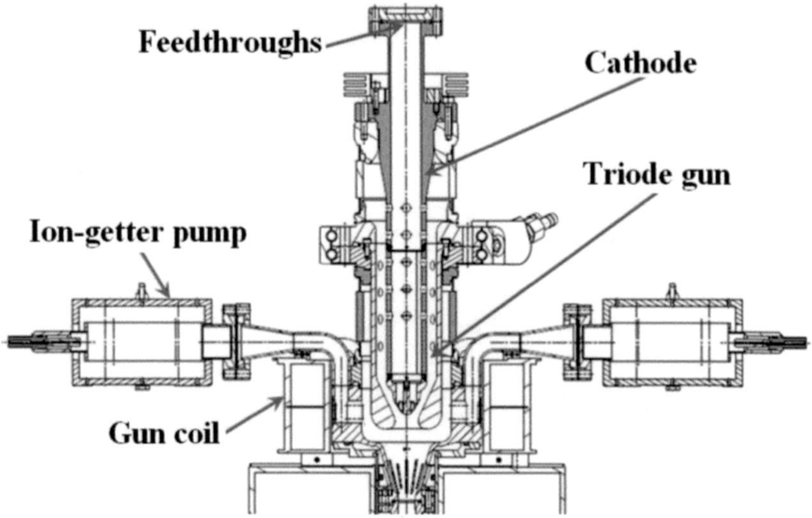

Figure 6.2: Configuration of triode magnetron injection electron gun

In Fig. 6.3 the cathode is shown which consists of cathode head, cathode stem, ConFlat (CF) flange for vacuum seal, and feed-through flange. For the initial experiments presented in this work, an M-type tungsten dispenser emitter has been used with a 3-pin feed-through flange. This 3-pin feed-through flange has been used to provide the cathode voltage and the power for filament heating.

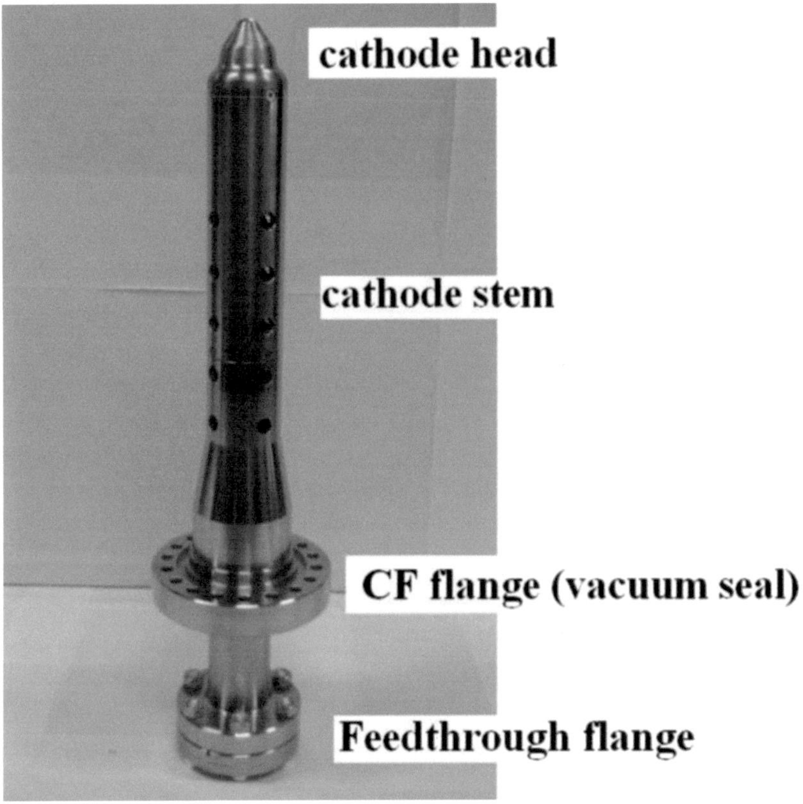

Figure 6.3: Cathode assembly

The emitter and the cathode head were manufactured and assembled by CCR in collaboration with Semicon company in the USA. Figure 6.4 shows a magnified photo of the cathode head where good precision of the assembly and the alignment of such parts as the cathode nose, emitter ring and back focus electrode can be seen. Before connection to the cathode stem, the cathode head has been tested in a vacuum bell jar (Fig. 6.5). The emitter has been heated up to 975 °C which is lower than its activation temperature (T = 1150 °C) but still enough to analyse the thermal behaviour of the different parts of the cathode head during heating. Measurement of the temperature distribution along the azimuth of the heated emitter ring has been performed with a pyrometer and showed that the deviation of the temperature less than 10 °C which is low enough to provide uniform emission of the electron beam during operation. Also from Fig. 6.5 it can be seen that the two adjacent parts to the emitter ring, the cathode nose and the back focus electrode are much lower in temperature than the emitter, which has been achieved by providing two thermal gaps. The first thermal gap is 0.12 mm wide and located between the emitter and cathode nose, the second one is 0.08 mm between the emitter and the back focus electrode. When the emitter is at room temperature, the gap between emitter and cathode nose has to be larger than the gap between emitter and the back focus electrode, because the emitter with its support structures extends and moves towards the cathode nose during heating, thereby decreasing the gap between the cathode nose and the emitter. The thermal design of the cathode head has been performed by CCR. One of the main goals of the thermal design was to avoid possible contact between the emitter, the cathode nose and the back focus electrode when the emitter is heated up to the operating temperature. At the same time, the thermal gaps should be small enough in order not to disturb the electric field distribution at the cathode surface.

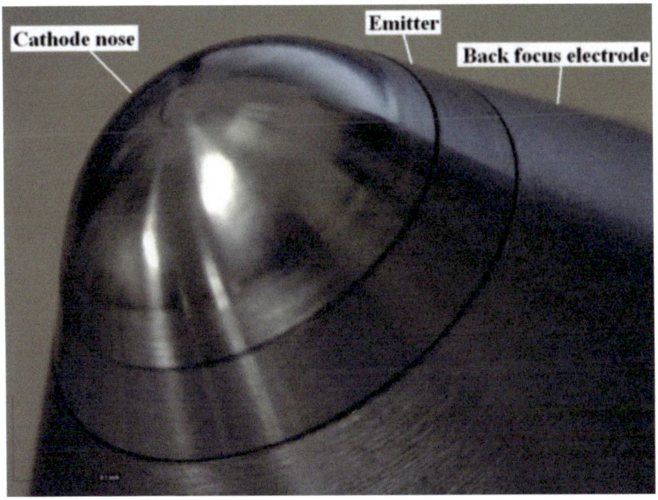

Figure 6.4: Cathode head

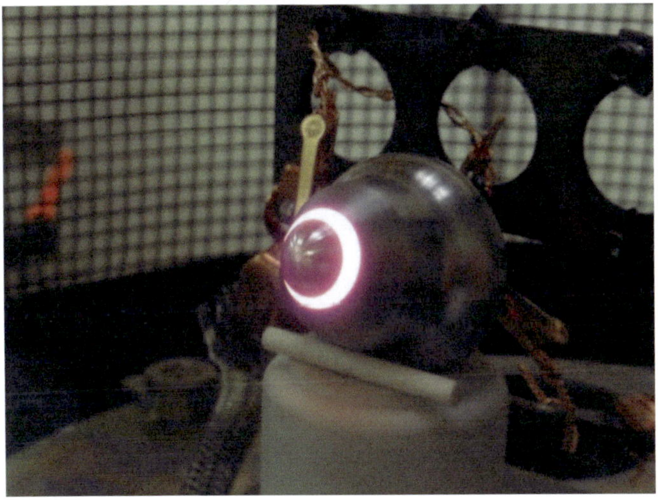

Figure 6.5: Cathode head with heated emitter inside a vacuum bell jar

The assembled triode electron gun is shown in Fig. 6.6. It includes cathode, modulation anode (copper part which can be seen at the top), first isolation ceramics with flanges between cathode and modulation anode and second isolation ceramics between modulation anode and anode.

Figure 6.6: Assembled triode electron gun with modulation anode at the top

6.3 Anode and beam tunnel

Figure 6.7 shows the part of the gyrotron which includes the anode, the beam tunnel and the stainless steel pipes for water cooling and vacuum systems. Preliminary drawings for the part shown in Fig. 6.7 have been done by the author.

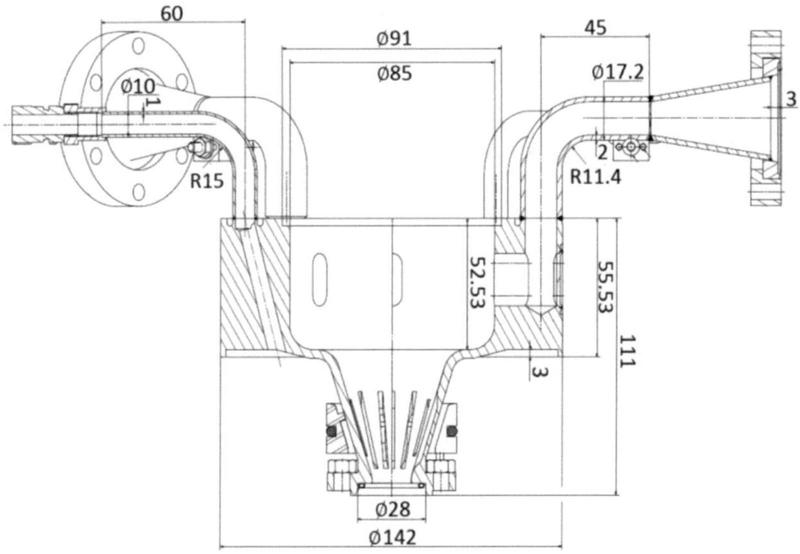

Figure 6.7: Cross section of the anode with the beam tunnel close to the cut-off section of the cavity

In order to avoid additional flanges or brazing connections, the anode and beam tunnel have been manufactured from one piece of stainless steel. Due to the requirement to fit into the electron gun coil (shown in Fig. 6.1), the vacuum and cooling pipes are connected from the top of the flange. There are four vacuum pipes: two pipes have been used for the ion-getter pumps, one pipe for the vacuum valve and one pipe as a spare for possible connection of an additional ion getter pump if it should be necessary.

Figure 6.7 shows a cross-section of the anode with the beam tunnel. The beam tunnel consists of a stainless steel cone with irregular groves in the wall. There are 12 cuts, which are distributed along the azimuth of the cone and have a depth of 2 mm, a width 1.8 mm and lengths from 23 mm to 32 mm. In the beam tunnel section parasitic modes can be excited because the electron beam in this region already has a relatively high pitch factor. These oscillations can decrease the electron beam quality and thus the efficiency of the interaction of the main mode and the electron beam in the gyrotron cavity. Such a construction of the beam tunnel made it possible to avoid excitation of parasitic oscillations before cavity due to interruption of the wall currents by the cuts and provided single mode generation in the cavity. Figure 6.8 shows a photo of the anode, beam tunnel, and the vacuum pumping and cooling water pipes.

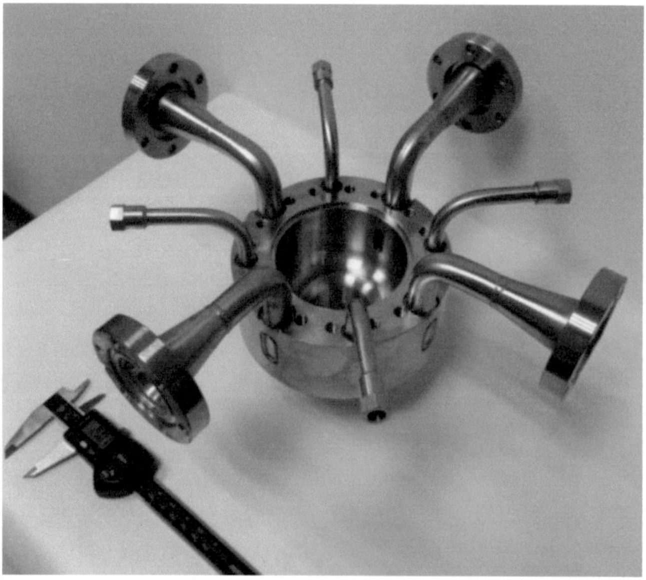

Figure 6.8: Photo of the anode with beam tunnel and cooling and vacuum pumping pipes

6.4 Cavity

Figure 6.9 shows a drawing of the cavity. It has been produced from Glidcop AL-15 material (Appendix A.2), which is with alumina dispersion strengthened copper that has very similar electrical and thermal conductivities to pure copper, but at the same time it is much stronger in terms of mechanical and thermal properties. The main material properties for different Glidcop composites can be found in Appendix A.2. Glidcop can sustain more cooling-heating cycles than oxygen free high conductivity (OFHC) copper, which is important for experimental investigation of the gyrotron in different regimes, with high thermal loads. One of the main reasons to use Glidcop material for this gyrotron is that it is strong enough to allow the use of flanges. As can be seen from Fig. 6.10, the cavity body and flanges at both ends of the cavity are produced from one piece of Glidcop material.

This has several advantages: first it eliminates the necessity to braze these two flanges; this increases the overall strength of the part and decreases the possibility of vacuum leaks. Second, manufacturing of the cavity and flanges from one piece of Glidcop material allows to achieve higher accuracy of the alignment of the cavity body and flanges.

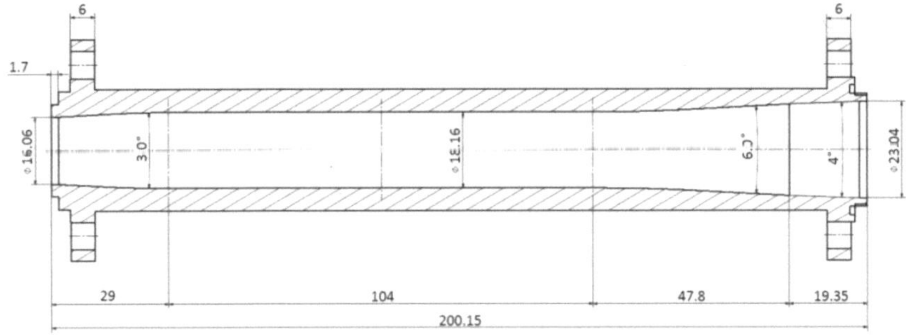

Figure 6.9: Technical drawing of cavity and up-taper

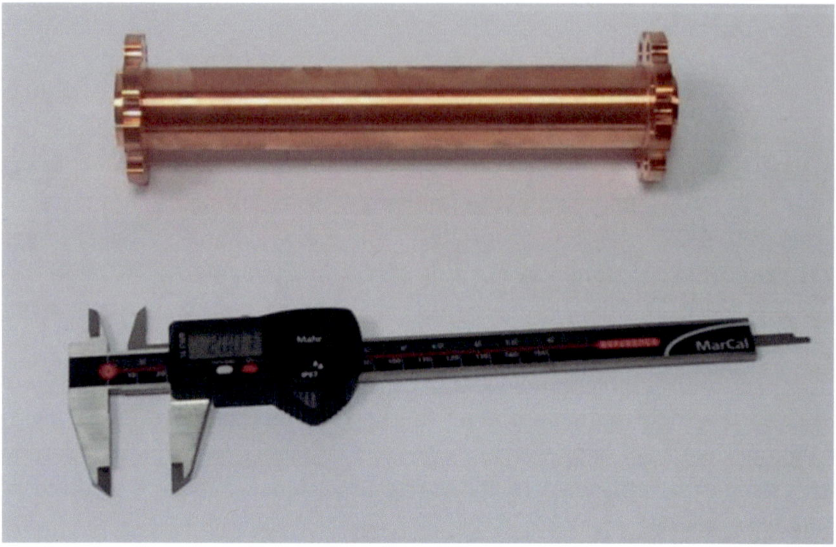

Figure 6.10: Photo of cavity and up-taper

6.5 Collector

The technical drawing and a photo of the collector which is connected to the output window structure are shown in Fig. 6.11. Preliminary drawings for the collector design have been done by the author. The collector has been manufactured from three parts due to its length (530 mm) and the difficulty to manufacture such a long and small in diameter part from one piece of copper. The inner collector diameter at the output section is 40 mm. From the cavity side it has an approximately 207 mm long up-taper (see Section 4) with a stainless steel flange (collector input diameter D = 23.04 mm) which is used to connect the cavity to the collector part through a flange connection with a Helicoflex type gasket. As it was shown by optimization in Section 4, the collector up-taper consists of two sections, one with an angle 4.96° that provides the area where the electron beam is absorbed, and the other with an angle 4.91° where no electrons are absorbed. This taper section has been used only as a smooth transition between the collector and the cavity. From the window side the collector has a stainless steel sleeve which is used to connect the window part by welding to another stainless steel sleeve. The collector is manufactured from copper in order to avoid possible problems with brazing of the flanges and for good heat conduction.

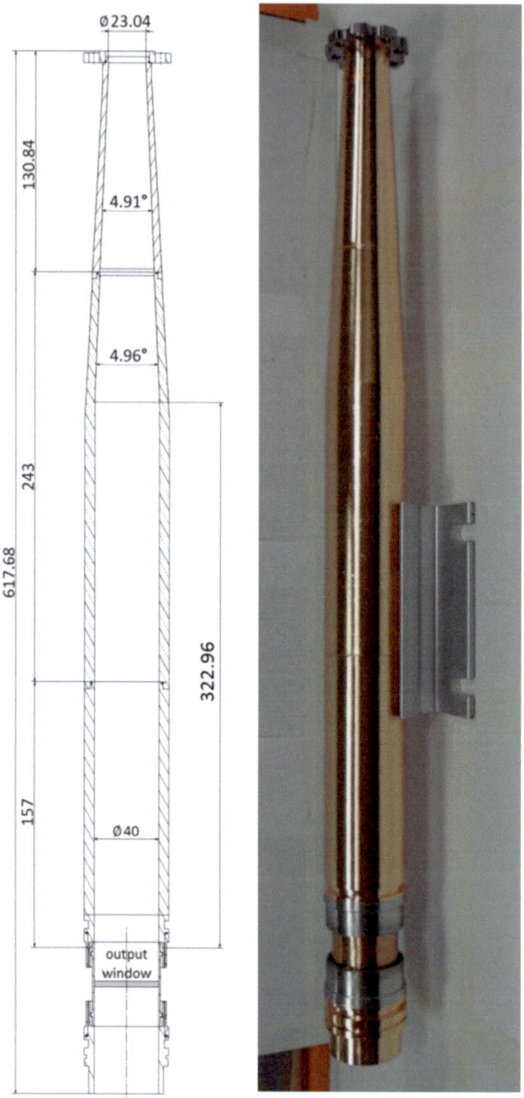

Figure 6.11: Drawing and photo of the collector connected to the window assembly

6.6 Output window

Figure 6.12 shows a photo of the output window brazed into the output waveguide, and two stainless steel sleeves which are brazed to the waveguide. Figure 6.13 shows a 3D drawing of the window assembly connected to the collector and output waveguide. Preliminary drawings of the window assembly have been done by the author. The design of the window part has been done in such a way that the waveguide has no interruptions; this was achieved by brazing the alumina AL-995 window disk inside the waveguide. This construction allows one to avoid reflection and unwanted conversion of the main mode $TE_{1,2}$ into other spurious modes. Such a construction of the window unit improves the overall efficiency of the gyrotron.

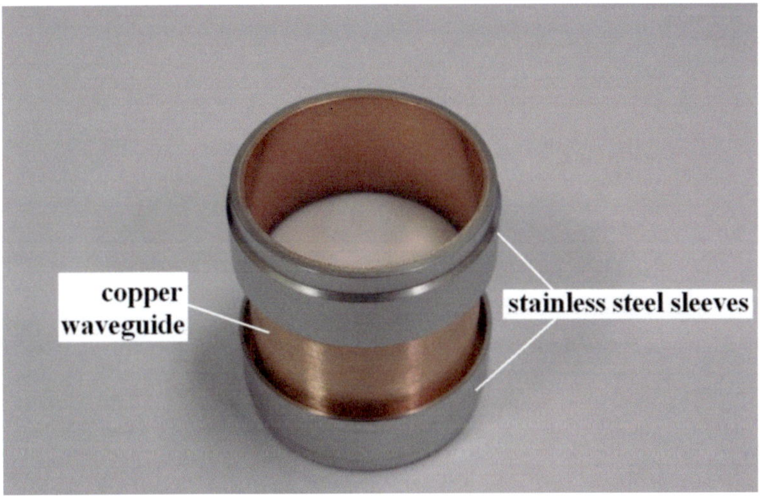

Figure 6.12: Photo of the window assembly

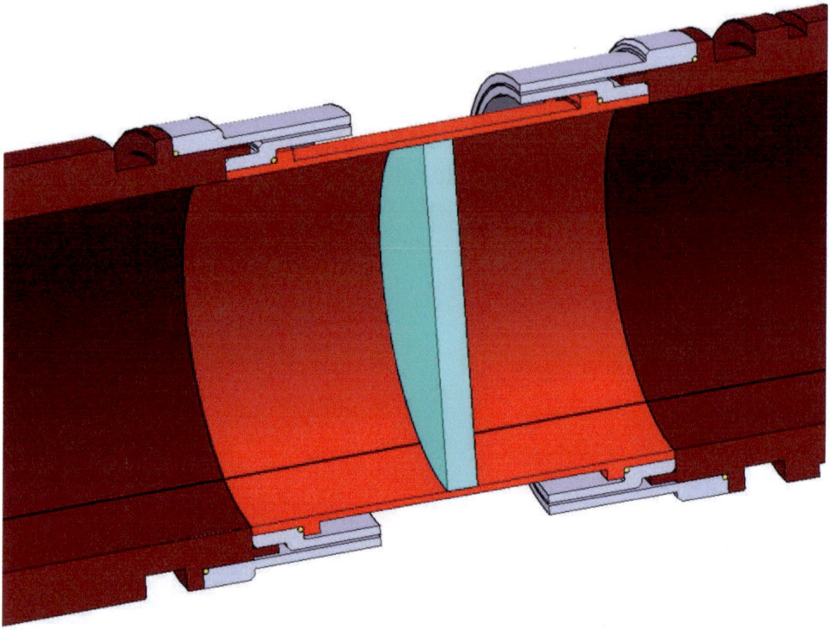

Figure 6.13: Drawing of the window assembly connected to the collector and output waveguide

Figure 6.14 shows a 3D drawing of the window assembly connected to the gyrotron and installed with the water cooling system. The window waveguide where the window disk has been brazed has its own water cooling loop separated from the water loop of the cavity and collector. This separation of the cooling loops allows to control pressure and water flow independently from other cooling loops which is very important because the window waveguide has relatively thin walls (thickness = 0.8 mm) and high water pressure may cause window disk cracking.

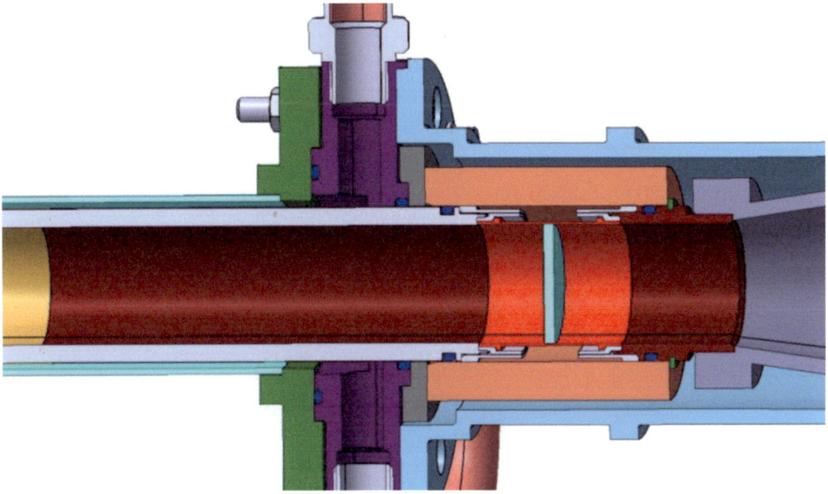

Figure 6.14: Drawing of the end of the gyrotron tube, including part of the collector, window assembly, part of the output cone, and cooling system

6.7 Bake-out procedure of the gyrotron

The gyrotron has to be prepared for operation in a CW regime. This means that the tube has to be baked-out. The bake-out procedure that has been set-up for this gyrotron is based on the procedure that is used for gyrotrons and klystrons produced by THALES and is described in [Lie10]. The maximum temperature that is used for bake-out of a MW class fusion gyrotron is 490 °C, and the overall bake-out time is around 14 days. The bake-out temperature increase (ramp-up) should be done with a rate not faster than 15 °C/hour in order to provide uniform heating of all parts of the gyrotron; the same rate should be used for decreasing the temperature during cooling of the bake-out oven.

In a later stage of the bake-out it is also necessary to heat up the emitter to temperatures close to its activation temperature (t = 1150 °C) [Ive12]. This makes it possible to evaporate different contaminations that can be present on the emitter surface. At the same time heating of the emitter creates a uniform *BaO* layer on its surface which is needed for the good performance of the emitter during operation.

For the gyrotron that is designed in this dissertation, the bake-out procedure was modified. The maximum temperature has been lowered to 250 °C in order to decrease stresses on different gyrotron components, such as, for example, isolation ceramics at the electron gun region, output window and also flange connections with Helicoflex type gaskets. The decrease of the temperature by a factor of 2 should in general increase the bake-out time by a factor of 2 [Pio12], but taking into account the smaller volume of the gyrotron designed in this dissertation (compared to MW gyrotrons) this factor should be around 1.5, which gives approximately 21 days for a bake-out procedure. All the gyrotron parts have been designed in order to withstand the 250 °C temperature treatment.

Figure 6.15 shows a photo of the gyrotron standing inside the bake-out oven on a special support frame. The vacuum pumping system is located outside the oven and is connected to the gyrotron through the corrugated and therefore flexible metal pipe, which is also shown on the photo.

Several thermocouples are connected to the gyrotron in order to control the temperature during the bake-out process as well as to control possible temperature differences between different parts of the gyrotron. The bake-out procedure has been controlled and performed by a special control system, which can be programmed to perform different bake-out scenarios. Figure 6.16 shows the dependence of the tube vacuum on time during bake-out.

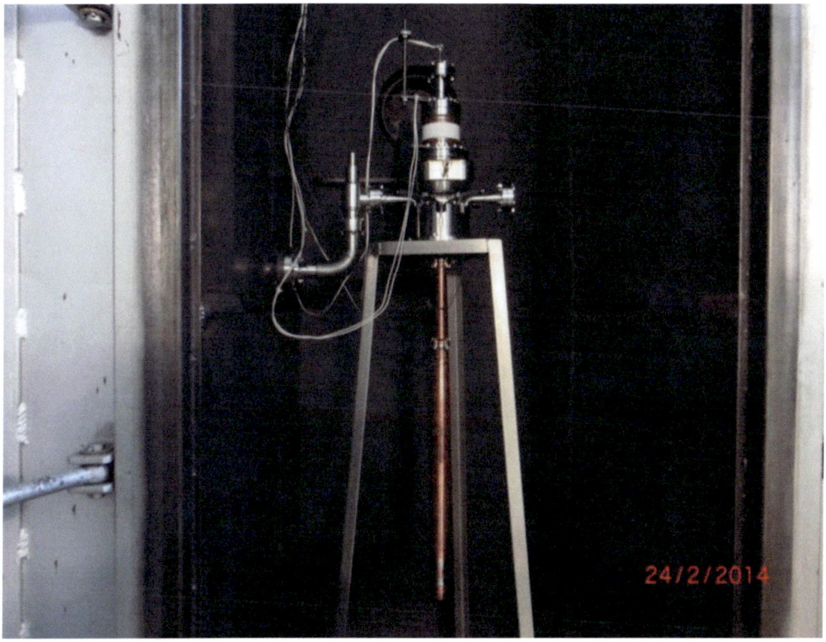

Figure 6.15: Photo of the gyrotron installed for the bake-out procedure inside the oven (at IHM)

The bake-out procedure consisted of three main phases. The first phase with a duration of 7 days is a step-wise increase of the temperature from room temperature to the maximum temperature which has been set to 250 °C. The second phase has been performed with a constant temperature (T = 250 °C) and lasted for 11 days. Finally, the third phase is the decreasing of the temperature from 250 °C to room temperature (T = 25 °C). It took 2 days.

The control system included several interlocks whose task was to react to different events during the bake-out procedure. One of the main events is an increase of the tube pressure during ramp-up of the temperature. As demonstrated in Fig. 6.16 the pressure did not exceed the level

$P_1 = 1 \cdot 10^{-6}$ mbar. This value has been set as the maximum value, and if the pressure in the gyrotron exceeds this value, the control unit of the oven stops increasing its temperature and goes to a regime of constant temperature. During the regime of constant temperature in most cases the vacuum improves (pressure decreases) and after crossing another vacuum interlock value which was set to $P_2 = 5 \cdot 10^{-7}$ mbar, the temperature was again allowed to increase with the same slope that was used before the first interlock (for P_1). A third vacuum interlock was set to $P_3 = 5 \cdot 10^{-6}$ mbar. If the pressure exceeds this value (P_3), the temperature decreases until the vacuum improves to values smaller than P_2. A fourth vacuum interlock has been set to $P_4 = 5 \cdot 10^{-5}$ mbar. When the pressure increases above this value, the bake-out procedure stops immediately and the temperature decreases much faster than in the case of exceeding the value P_3 as was described above. It can be seen from Fig. 6.16 that the tube pressure several times reached the value $P_1 = 1 \cdot 10^{-6}$ mbar and the temperature went to a constant value for some time. But after recovery of the vacuum to values smaller than $P_2 = 5 \cdot 10^{-7}$ mbar the increase of the temperature continued. During the second phase (constant temperature of T = 250 °C) one can see very smooth improvement of the vacuum conditions inside the tube which indicated proper outgassing of the internal surfaces of the gyrotron.

Heating of the emitter of the gyrotron has been performed at the end of the second phase as can be seen from Fig. 6.16 where three peaks are present. Heating of the emitter has been performed during three days, with a duration of 8 hours every day. The increase of the pressure during heating of the emitter indicated outgassing of contaminations at the emitter surface. The final tube pressure which has been achieved at the end of the bake-out process was $P = 1 \cdot 10^{-8}$ mbar. Separation of the gyrotron from the pumping system (which was located outside the oven, as has been mentioned) has been done by using the special valve which is shown in Fig. 6.15. After closing the valve, magnets have been attached to the two ion-getter pumps. After switching on the ion getter pumps the vacuum improved even further to $P = 1 \cdot 10^{-9}$ mbar. The main task of the ion-getter pumps is to provide good vacuum (P < 1 · 10^{-7} mbar) during CW operation of the gyrotron.

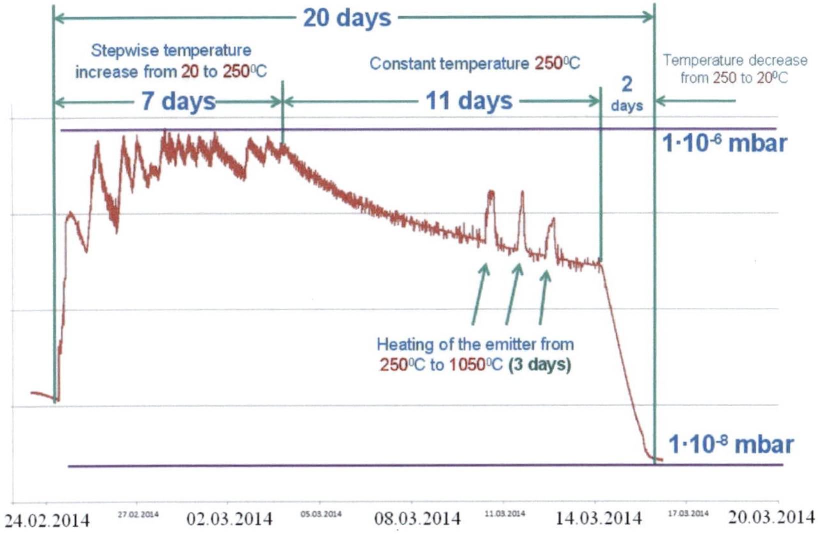

Figure 6.16: Gyrotron vacuum as a function of time during the bake-out procedure

6.8 Test stand and water cooling system

Figures 6.17 and 6.18 show photos of the assembled gyrotron after the bake-out procedure and of the test stand with the installed gyrotron. The preliminary design of the test stand was done by the author. The test stand has been constructed using standard aluminium profiles which allows easy and fast modifications if necessary. The test stand consists of three frames: the main frame and two additional frames connected at the back of the main frame, which are shown in Figs. 6.19 and 6.20. The main frame as can be seen from Fig. 6.18, includes the gyrotron, the magnet system, the calorimeter, the pipes of the cooling system, flow meters, as well as temperature and pressure sensors. Figure 6.19 shows the additional frame where the power supply for the filament (heater of the emitter) is installed. In Fig. 6.20 the control system of the gyrotron test stand is shown which is based on a Siemens Simatic S7 control system. A similar system is used for the control of the test stand for MW-class fusion gyrotrons at IHM [DBC+99].

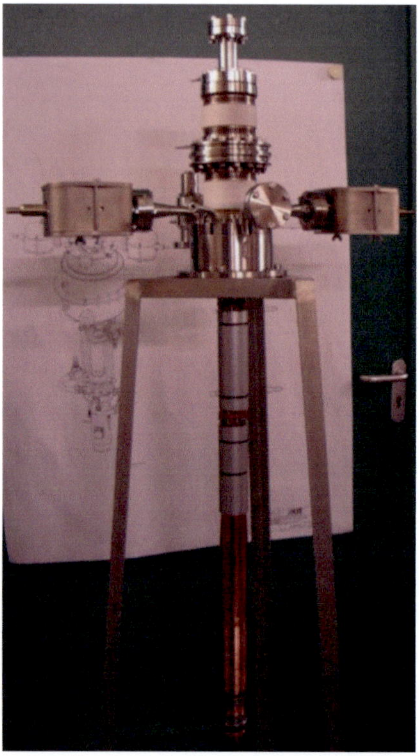

Figure 6.17: Photo of the assembled gyrotron

Figure 6.18: Gyrotron with calorimeter installed in the test stand

Figure 6.19: Gyrotron filament power supply

Figure 6.20: Gyrotron Simatic S7 control system

6.9 Cooling water system

In Fig. 6.21 the scheme of the cooling water system of the different gyrotron components is shown. The preliminary design of the configuration of this system has been done by the author. The cooling water system includes several water loops: anode/cavity, collector, output window, modulation anode and calorimeter. The flow rates of the different water loops are summarized in Table 6.1. The water loop of the collector is a continuation of the anode/cavity water loop with four additional input pipes connected at the beginning of the collector region (Fig. 6.22). As is the case for the output window and the modulation anode, the calorimeter also has a separate water loop. It is done in order to avoid possible influences from other water loops and to provide precise power measurements. The input water temperature has been set to 20 °C.

Table 6.1: Flow rates of the different flows loops of the cooling system

Water loop	Flow [l/min]	Output temperature [°C]	Maximum pressure, [bar]	Pressure drop [bar]
Mod. anode	0.2 – 0.4	50 – 35	8	2
Anode/cavity	4.3 – 8.6	50 – 35	8	2.5
Collector	65 – 130	50 – 35	8	2.5
Output window	0.3 – 0.6	50 – 35	3	0.5
Calorimeter	7.5 – 15	50 – 35	8	2

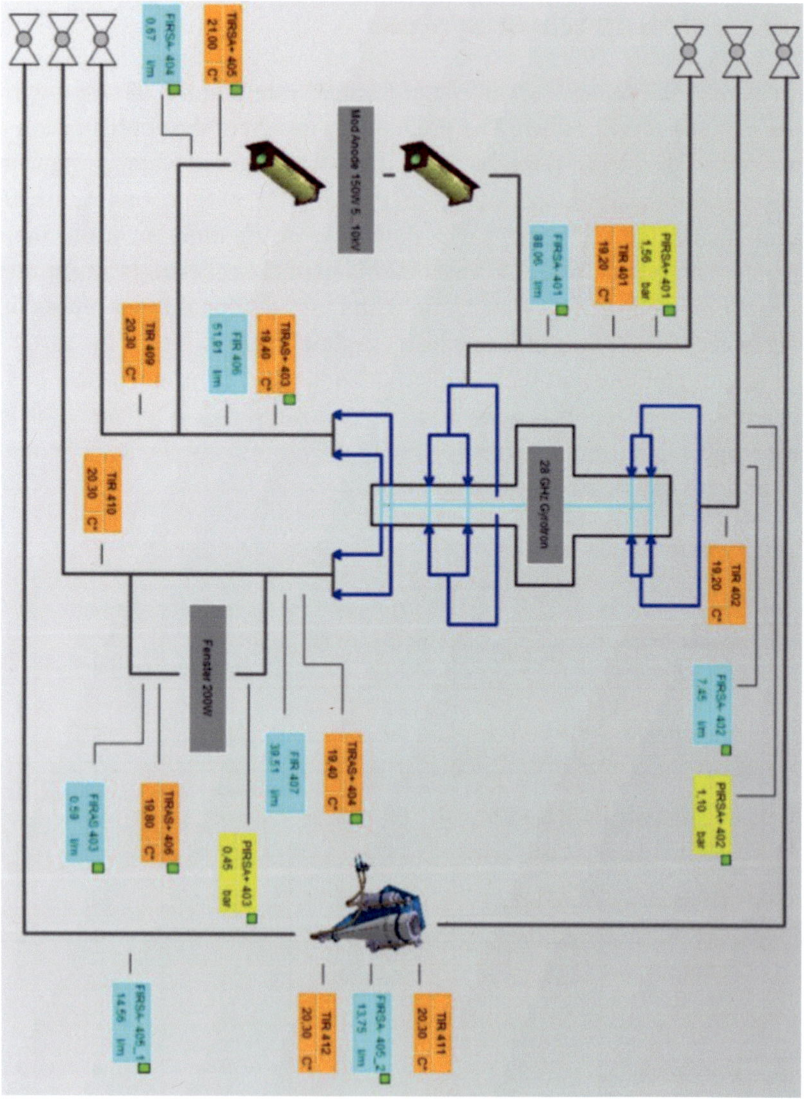

Figure 6.21: Cooling water system of the gyrotron test stand

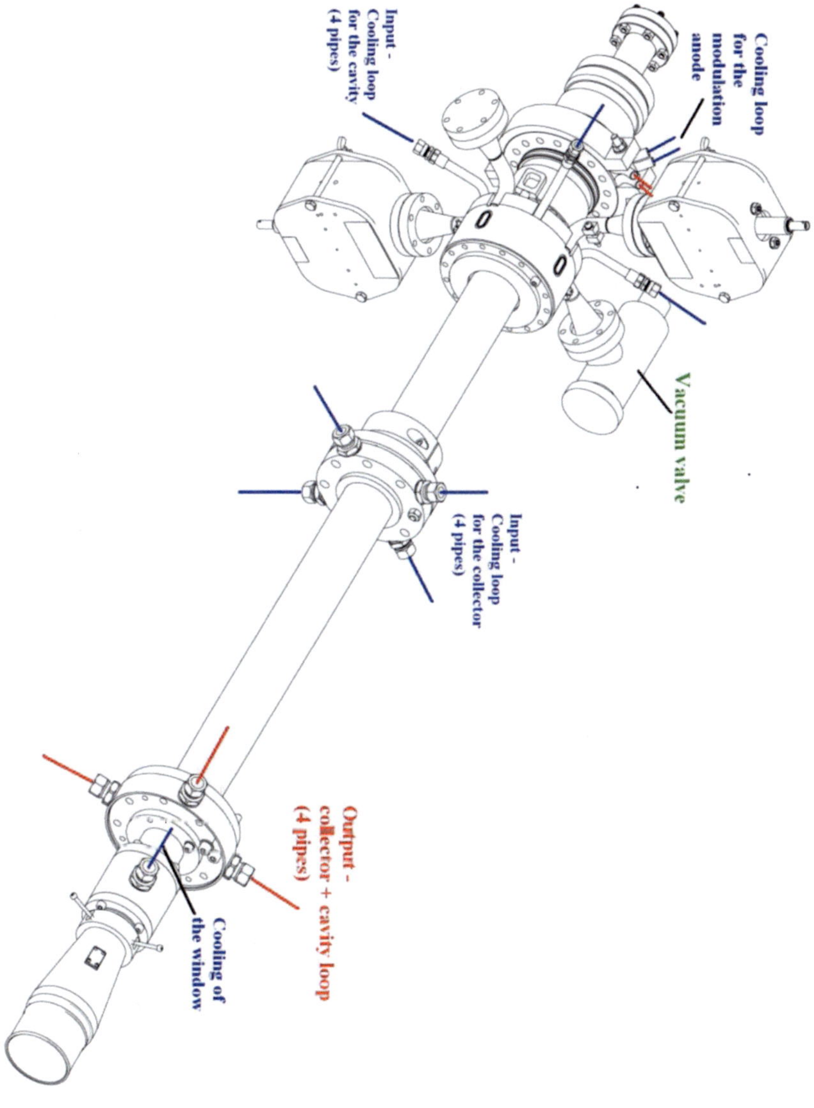

Figure 6.22: 3D-drawing of the gyrotron together with mounted cooling water supply system

7 Experimental investigations

After the bake-out procedure had been performed, the gyrotron was installed in the test stand and the conditioning process was performed. This conditioning process included reaching the nominal electron beam current for the emitter. After this process of cleaning the internal surfaces of the gyrotron from contaminations and the presence of a thin layer of gases with the help of the electron beam and electromagnetic waves generated in the cavity.

The conditioning procedure started from low electron beam currents (I < 0.01 A) and low voltages (U < 1 kV) with a pulse length t < 1 sec. The series of pulses were repeated several times with the same beam parameters in order to improve the vacuum and guarantee that a low vacuum reaction during pulses would be achieved. When the vacuum was improved, the electron beam current, cathode voltage and pulse length were increased step by step. The conditioning procedure was performed over a period of two weeks and made it possible to achieve good vacuum conditions of the internal surfaces of the gyrotron and the nominal electron beam current with nominal cathode and modulation anode voltage.

An example of a pulse is shown in Fig. 7.1 where the electron beam current and cathode voltage are plotted. The pulse duration was set to 30 seconds. During the pulse the heating power of the emitter (filament) was kept constant. It can be observed that the electron beam current decreases. This happens due to the cooling effect of the emitter surface when electrons leave it. The electron current that flows through the cathode has electrons that enter the cathode from the "cold" side where the cathode is at room temperature and leave the cathode from the emitter surface which is heated to 1150 °C. This means that when the electron current starts to flow through the cathode (after the start of the pulse), electrons start to leave the heated emitter surface, decreasing its temperature, which decreases the electron beam current. After some time an equilibrium temperature is reached and the current stabilizes. This happens when the energy that electrons have taken from the emitter surface is compensated by the heating of the filament (Fig. 7.2).

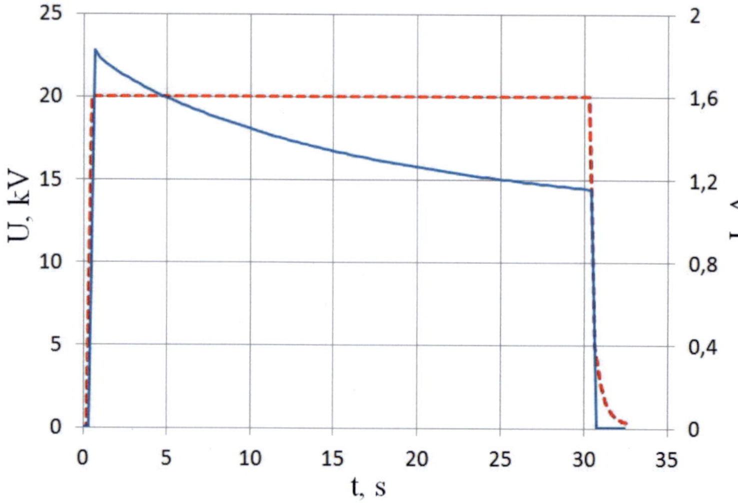

Figure 7.1: Electron beam current (solid curve) and cathode voltage (dashed curve) as a function of time, with constant filament heating. The pulse length is 30 sec

Longer pulses showed that after a pulse duration of around 1-2 min the electron beam current reaches a saturation level. Figure 7.2 presents a pulse with the length t = 120 seconds. It can be seen that after approximately 80 seconds the decrease of the electron beam current becomes very small compared to the rapid drop that occurs during the first 20 seconds.

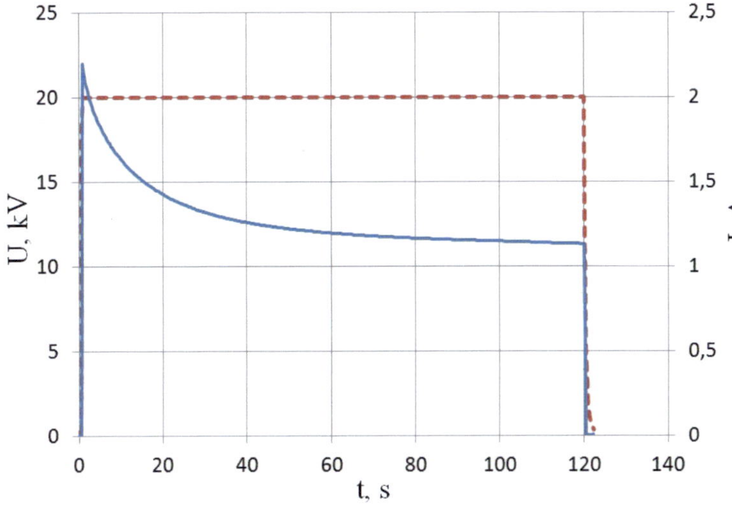

Figure 7.2: Electron beam current (solid curve) and cathode voltage (dashed curve) versus time, with constant filament heating. The pulse length is 120 sec

In order to keep the electron beam current at its original value it is necessary to increase the filament heating (by increasing (boosting) the filament current) in order to compensate the cooling effect of the emitter surface. An example of a pulse where an increase (compensation) of the filament heating has been performed is shown in Fig. 7.3. It can be seen that in the first several seconds the current decreases but then the increase of the filament current compensates cooling of the emitter during the pulse. Then the beam current increases to the nominal value and reaches saturation. This pulse shows just one example of how the current drop can be almost compensated. A system with a feedback loop is needed for such experiments.

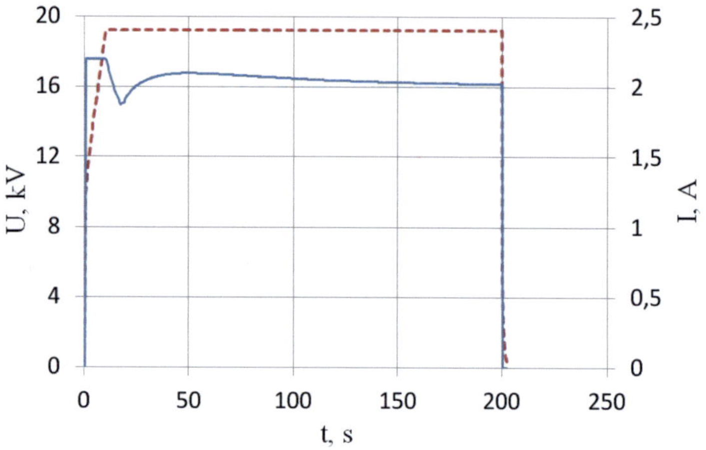

Figure 7.3: Electron beam current (solid curve) and cathode voltage (dashed curve) versus time, with boosting of the filament heating

7.1 Experimental results

An example of a 17 kW, 5000 s long pulse is shown in Fig. 7.4 where the vacuum inside the tube (blue curve), measured by one of the ion-getter pumps, output power (brown curve) measured by the calorimeter, cathode voltage (blue curve) and electron beam current (green curve) are plotted. The voltage is kept stable automatically by the high-voltage power supply. At the beginning of the pulse it can be seen that stabilization of the current has been performed; the procedure has been described in a previous section. The internal tube pressure during the pulse is stable and does not exceed the limiting value of $5 \cdot 10^{-7}$ mbar which was set as a maximum value for the interlock system. The frequency of the output wave has been measured by a spectrum analyser and is shown in Fig. 7.5. The measured frequency f = 28.03 GHz agrees very well with the theoretical simulations.

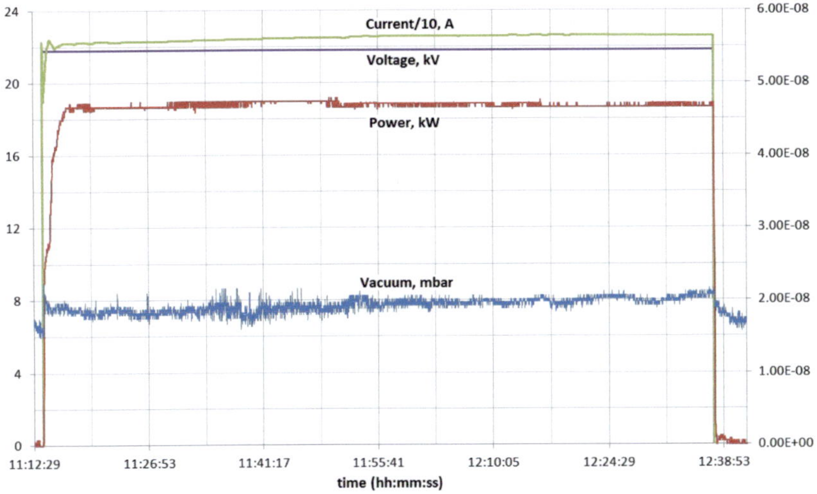

Figure 7.4: 17 kW, CW gyrotron operation (pulse length 5000 sec)

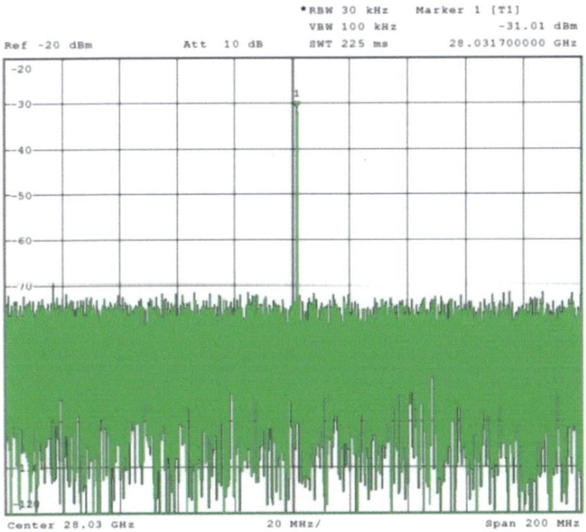

Figure 7.5: Spectrum of the RF output wave. The gyrotron is operated in the main second harmonic $TE_{1,2}$ mode

Different regimes of gyrotron operation have been studied. The additional coil (coil# 4) has been used to optimize the magnetic field profile in the cavity in order to reach a record efficiency of 43% for CW operation of the second harmonic 28 GHz gyrotrons. The optimum current of the coil #4 was 1.2 A. As it can be seen from the list below, regimes with low emitter current density have been also demonstrated, these are suitable for material sintering applications [BEG+04, LFT+99, TLST04, LHJ+04]

The main regimes that have been studied are listed below:

1) U_{beam} = 23.4 kV, I_{beam} = 2.23 A, P_{RF} = 22.5 kW, η = 43 % - high efficiency, high cathode voltage operation.
2) U_{beam} = 21.8 kV, I_{beam} = 2.23 A, P_{RF} = 19 kW, η = 39 % - high electron beam current operation.
3) U_{beam} = 20.6 kV, I_{beam} = 1.47 A, P_{RF} = 11.4 kW, η = 37 % - average electron beam current regime.
4) U_{beam} = 20.6 kV, I_{beam} = 1.02 A, P_{RF} = 7.6 kW, η = 36 % - low electron beam current regime, with an emitter current density of 1.2 A/cm^2 for long lifetime operation of the M-type emitter.

In addition to operation with constant power in CW operation as is shown in Fig. 7.4, regimes with smooth (step-wise) change of the output power during the pulse are important for material sintering applications [BEG+04, LFT+99, TLST04, LHJ+04, RLFT00]. Several approaches can be used in order to change power during operation. The first method is to vary the output power by changing the main magnetic field via variation of the current in the main coils. This is demonstrated in Fig. 7.6, where the output power is decreased from 22.5 kW to 1.6 kW by step-wise decrease of the cavity magnetic field. Each step has a time duration of around 30 seconds and represents one magnitude of the cavity magnetic field. It can be seen from Fig. 7.6 that stable generation of the main mode $TE_{1,2}$ can be achieved over a wide range of output powers. The maximum power 22.5 kW is generated in a stable regime for more than 1000 seconds.

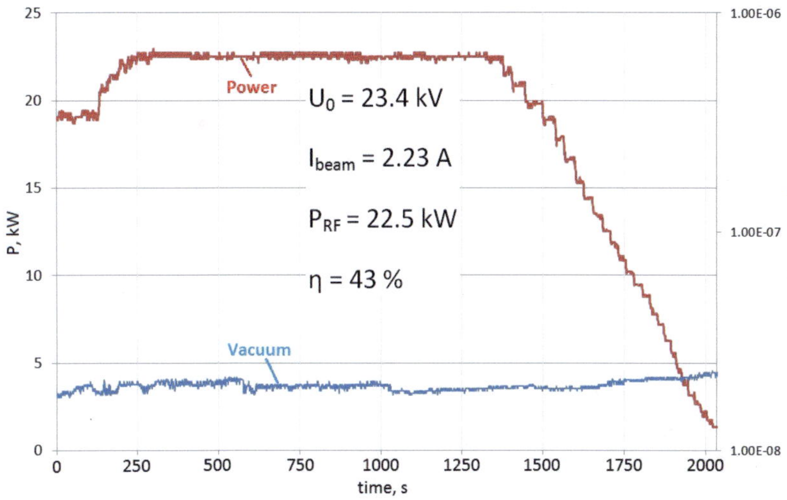

Figure 7.6: Stepwise change of output power by variation of the cavity magnetic field

The second approach for changing the output power is by changing the cathode voltage. In order to keep the pitch factor constant, the modulation anode voltage has to be varied accordingly. This approach allows fast variation of the output power, because cathode and modulation anode power supplies make it possible to change the voltages much faster than the power supply for the main magnet.

8 Comparison of the experimental and theoretical results

In Section 7, the experimental results which were acquired during experimental studies of the gyrotron have been described and summarized. In this section comparison of the experimental results and simulation results will be shown.

At first the high efficiency high voltage operating point which was shown in Section 7 (Fig. 7.6) will be compared with simulation results. In Fig. 7.6 in addition to the demonstration of the stable generation of the RF power of 22.5 kW during 1000 seconds, a step-wise change (increase) of the magnetic field was also demonstrated where each step was approximately 30 seconds long. This made it possible to vary the output power from 22.5 kW to 1.6 kW.

In simulations, sweeps over the magnetic field with the EURIDICE code were performed using a step-wise approach. In this case a certain magnetic field was fixed for a certain amount of time, then step wise switched to a smaller or higher magnetic field with a predefined step. In order to do such a step-wise sweep, the main cavity mode first had to be excited in the standard start-up scenario (example in Fig. 2.10). This point is shown as time = 0 ns in Fig. 8.1 and then the magnetic field was changed stepwise. In simulations the same electron beam parameters were used as in the experiment: U_{beam} = 23.4 kV, I_{beam} = 2.23 A, g = 1.8, Δg = 8 %, and $I_{coil\#4}$ = 1.2 A. The pitch factor and pitch factor spread were calculated employing the ESRAY code. Figure 8.1 shows a comparison between experimental results and simulation results. In the simulations each step over the magnetic field is equal to 0.003 T.

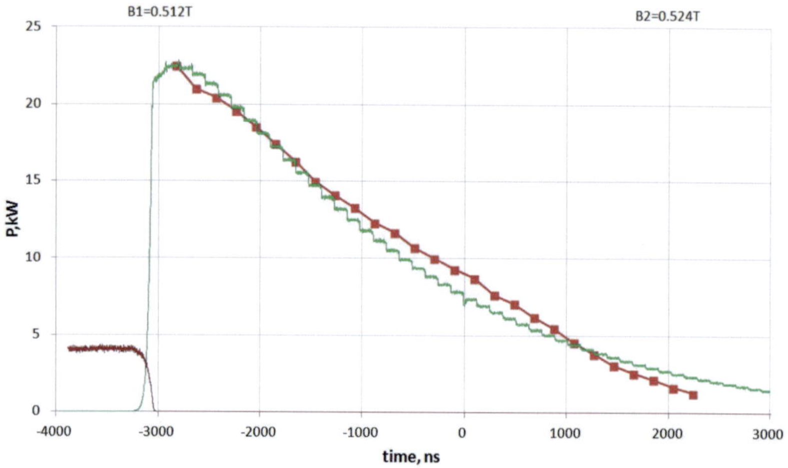

Figure 8.1: Comparison of experimental results (squares) with simulations for the high efficiency regime U_{beam} = 23.4 kV, I_{beam} = 2.23 A, P_{RF} = 22.5 kW, η = 43 %. The time steps correspond to stepwise changes of the maximum cavity magnetic field from 0.512 to 0.524 T

Analysing Fig. 8.1 one can conclude that the results of simulations agree very well with the results obtained from measurements. Both experiments and simulations showed that the main mode $TE_{1,2}$ is suppressed at cavity magnetic fields lower than 0.512 T by the competing backward wave mode $TE_{1,1}$ at the first harmonic of the cyclotron frequency. Its power radiated to the direction of the electron gun region amounts to approximately 4 kW as shown by the other red curve in Fig. 8.1.

In the following, one can see additional comparisons of the data achieved experimentally and in simulations for the following electron beam parameters: U_{beam} = 20.6 kV, I_{beam} = 2.08 A, Δg = 8 %, and $I_{coil\#4}$ = 1.2 A. Figure 8.2 shows the similarities between simulation and experiment.

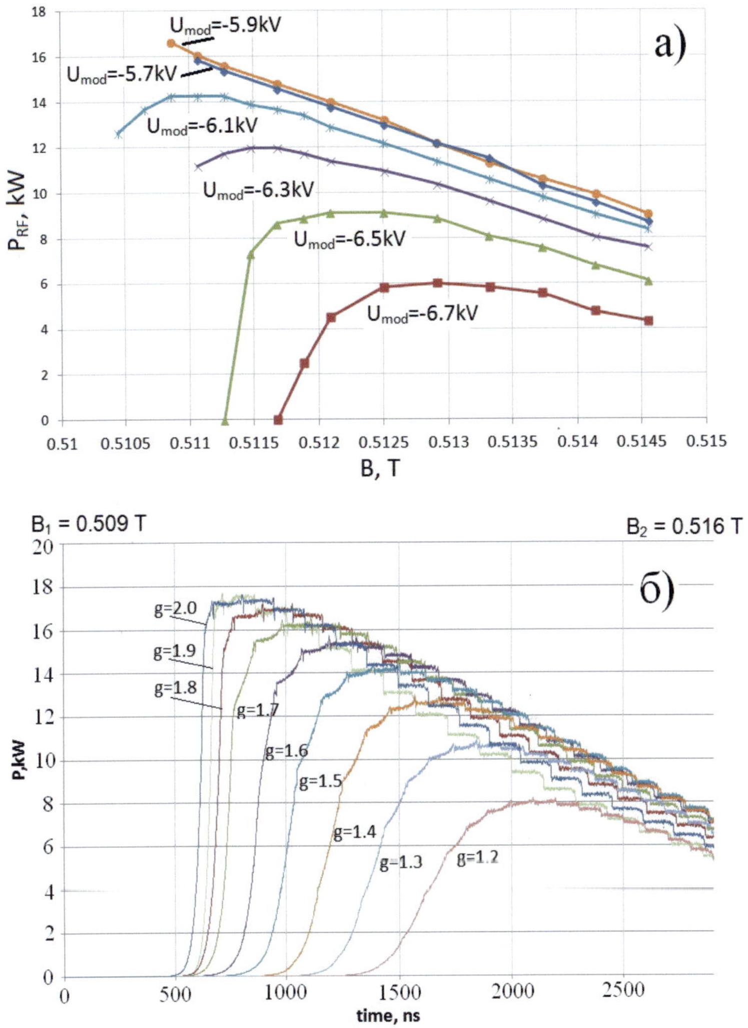

Figure 8.2: RF output power of the gyrotron as a function of magnetic field for different pitch factors achieved in (a) the experiment where the pitch factor was varied by variation of the modulation anode voltage and (b) simulated with the EURIDICE code

Several effects can be observed, as predicted by the simulations, the experiment reveals a saturation of the power after increasing the pitch factor above a certain level. In Fig. 8.2 U_{cath} = -20.6 kV, and U_{mod} is the voltage of the modulation anode.

Several other details which were observed in experiments were also found in simulations. In Fig. 8.3 the same electron beam parameters have been used as in Fig. 8.2. One can see from Fig. 8.3 that if the gyrotron is operated at high pitch factors (g = 1.8), the sharp break of the generation of the main second harmonic cavity mode $TE_{1,2}$ at the low-field side leads to the excitation of the spurious backward wave mode $TE_{1,1}$ at the first harmonic of the electron cyclotron frequency. Whereas for smaller pitch factors around g = 1.2, the reduction of the main mode generation is much smoother and there is no excitation of any other mode, including the spurious $TE_{1,1}$ mode.

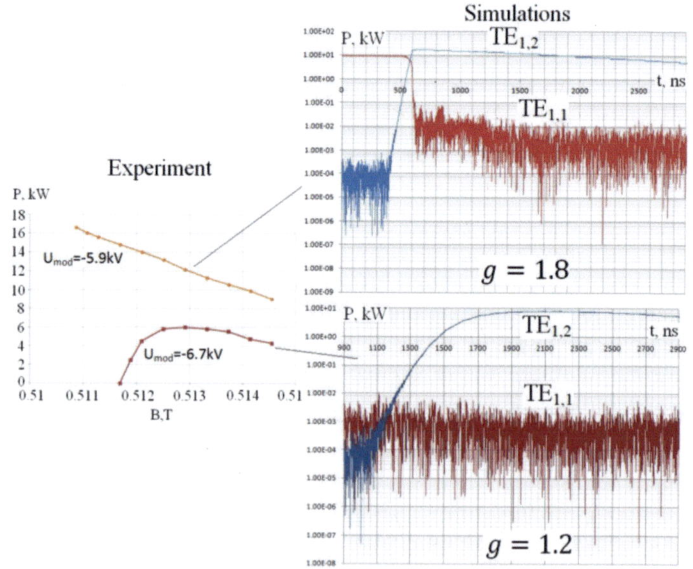

Figure 8.3: RF output power of the gyrotron as a function of magnetic field for different pitch factors achieved in (left) the experiment where the pitch factor was varied by variation of U_{mod} and (right) simulated with the EURIDICE code.

9 Conclusions and outlook

In this work the design and experimental investigations of a 20 kW / 28 GHz second harmonic gyrotron for evaluation of new emitter technologies were presented. The design phase of this gyrotron included numerical simulations of the cavity, the triode-type magnetron injection electron gun, the collector and the output window. Preliminary studies of different cavities that satisfy the gyrotron requirements made it possible to choose the most appropriate cavity with the main mode $TE_{1,2}$. This cavity mode has the weakest possible competition from the spurious mode $TE_{1,1}$ at the first electron cyclotron harmonic. In order to optimize the geometrical cavity profile, simulations with the non-stationary multimode code EURIDICE were performed. Optimization provided highly efficient, stable generation of the main second harmonic $TE_{1,2}$ mode for a wide range of pitch factors $1.1 < g < 1.8$ and a wide range of operating currents $0.5 < I_{beam} < 2.3$ A.

The next step in the design procedure was to optimize the three-electrode electron gun. A triode-type instead of a diode-type magnetron injection gun was chosen, in order to increase the range of operating parameters. The electron gun was simulated with the ESRAY code. A multi-parametric optimization was performed for a wide range of operating parameters. The main goal of the optimization was to minimize the pitch factor spread for a wide range of emitter current densities (0.5 A/cm^2 $< J_{em} <$ 2.5 A/cm^2) as well as pitch factors ($1.1 < g < 1.8$) and to provide a proper electron beam radius (R_{beam} = 3.13 mm) for stable single-mode operation of the main mode $TE_{1,2}$ at the second harmonic of the electron cyclotron frequency.

The third step of the design was optimization of the collector profile. The ESRAY code was used for these collector simulations. Multi parametric optimization of the collector geometry allowed to achieve reduction of the peak power loading from 500 W/cm^2 to 300 W/cm^2 for the beam parameters U_{beam} = 20 kV and I_{beam} = 2.2 A. For the maximum operating parameters that could be used in experimental investigations U_{beam} = 25 kV and I_{beam} = 2.3 A the peak power density is around 400 W/cm^2, which is still lower than the maximum allowed peak value of 500 W/cm^2.

In order to provide proper evaluation of new emitter technologies in future, the gyrotron has to fulfill specific technical requirements. The first requirement is that the gyrotron has to be able to operate in the CW regime. For this, a proper cooling system has been designed. The gyrotron was baked out during three weeks at a maximum temperature of 250 °C. This made it possible to remove all the residual gases and contaminations from the internal surfaces of the gyrotron. In addition to this, in order to provide good vacuum during operation, the gyrotron was equipped with two ion-getter pumps that could provide a vacuum down to $1 \cdot 10^{-7}$ mbar during operation. The second requirement is to provide a modular design of the gyrotron. The main components of the gyrotron are connected with flanges and special Helocoflex gaskets. This construction allows to perform bake-out of the gyrotron, which provides good vacuum, and at the same time the flanges allow fast exchange of the components of the gyrotron for future experimental studies of new emitter technologies as well as new configurations of gyrotron components.

Experimental investigations showed that the main $TE_{1,2}$ mode at the second harmonic of the cyclotron frequency can be excited over a wide range of operating parameters. A stable long pulse operation has been demonstrated for a wide range of output powers from 1.6 kW to 22.5 kW, with an internal tube pressure not exceeding $1 \cdot 10^{-6}$ mbar. Optimization of the electron gun made it possible to reach experimental operation with very high pitch factors for careful stability studies. High pitch factor regimes (g = 1.8) allowed to reach the world record efficiency of 43 % for CW second harmonic 28 – 30 GHz gyrotrons [Thu15].

P_{RF} [kW]	Efficiency [%]	Manufacturer
22.5	**43**	**KIT, Germany**
10	42	IAP RAS, Russia
10	38.7	Mitsubishi, Japan
13	33.6	CPI, USA

Experimental studies also showed that the gyrotron can be used for material processing applications. It can operate in stable CW regimes, where the output power can be smoothly changed over a wide range ($1.6 < P_{RF} < 22.5$ kW) by changing the strength of the main magnetic field or by modulation of the anode voltage, which is a very essential requirement for providing tailored heating profiles in material processing applications.

Comparison of the theoretical and experimental results showed good agreement. In simulations a maximum achievable efficiency of the main mode in a regime with high pitch factor strongly depends on competition with the first harmonic $TE_{1,1}$ backward wave, where as in the regimes with low pitch factors, excitation of the backward wave was not present. These effects have also been demonstrated in the experimental studies.

One of the most promising cathode technologies that can provide uniform distribution of the electron emission and a long lifetime is a CPR cathode. Up to now, such a CPR gyrotron cathode can be manufactured only from segments. Such a segmented cathode emitter permits one to study experimentally the influence of non-uniform emission on the operation of the electron gun and electron beam – RF wave interaction in the cavity.

The gyrotron designed, manufactured and tested in the scope of this dissertation allows one to perform a proper evaluation of new emitter technologies for future fusion gyrotrons. The construction of the gyrotron provides a fast and easy exchange of the emitter. A wide range of operating parameters of the electron gun and cavity makes it possible to study the emitter in different regimes.

List of figures

List of tables

Bibliography

[AB57] B.H. Alexander and R.W. Balluffi, Acta Met., vol. 5, no. 11, pages 666–677, 1957.

[AKK+99] R. Advanti, S. Korbly, K. Kreischer, M. Shapiro, and R. Temkin. Multi-megawatt gyrotrons for ECRH. In Proc. 24th Int. Conf. Infrared & Millimeter Waves, Monterey, CA, Sep. 1999.

[APIV12] K. A. Avramides, I. G. Pagonakis, C. T. Iatrou, and J. L. Vomvoridis. EURIDICE: A code-package for gyrotron interaction simulations and cavity design. In Proc. Eur. Phys. J. Web Conf., 2012, vol. 32, id 04016.

[BBL+69] R.J. Bondley, W.T. Boyd, R.G. Lock, T.J. Nall, and M.J. Slivka. High current density cathodes. US Army Technical Report ECOM-02289F, Jan. 1969.

[BEG+04] Yu. Bykov, A. Eremeev, M. Glyavin, V. Kholoptsev, A. Luchinin, I. Plotnikov, G. Denisov, A. Bogdashev, G. Kalynova, V. Semenov, and N. Zharova. 24–84-GHz Gyrotron Systems for Technological Microwave Applications. IEEE Trans. on Plasma Science, Vol. 32, No. 1, February 2004, pages 67-72.

[Ber11] M. H. Beringer. Design studies towards a 4MW 170 GHz coaxial-cavity gyrotron. Karlsruher Forschungsberichte aus dem Institut für Hochleistungsimpuls- und Mikrowellentechnik, Band 1, PhD thesis, Karlsruhe Institute of Technology (KIT), 2011. ISBN 978-3-86644-663-2.

[Bon69] R.J. Bondley. US Patent 3 434 812, March 1969.

[Bor91] E. Borie. Review of Gyrotron Theory. Kernforschungszentrum Karlsruhe, Karlsruhe, Germany, 1991.

[Chu04] K. R. Chu. The electron cyclotron maser. Rev. Mod. Phys., 76:489–540, May 2004.

[Cho14] A. R. Choudhury. Investigations of After Cavity Interaction in Gyrotrons Including the Effect of Non-uniform Magnetic Field. Karlsruher Forschungsberichte aus dem Institut für Hochleistungsimpuls- und Mikrowellentechnik, Band 4, PhD thesis, Karlsruhe Institute of Technology, Karlsruhe, 2014. ISBN 978-3-7315-0129-9.

[DBC+99] G. Dammertz, O. Braz, A. K. Chopra, K. Koppenburg, M. Kuntze, B. Piosczyk, and M. Thumm. Recent results of the 1MW, 140 GHz, $TE_{22,6}$-mode gyrotron. IEEE Transactions on Plasma Science, 27(2): 330–339, 1999.

[DBE+07] G. Denisov, Yu. Bykov, A. Eremeev, M. Glyavin, V. Kholoptsev, G. Kalynova, A. Luchinin, M. Morozkin, I. Plotnikov, D. Sobolev. High efficient gyrotron based systems for materials processing. Proc. 8th IEEE Int. Vacuum Electronics Conference (IVEC 2007), Kitakyushu, Japan, 433-434, 2007.

[DBG+08] G.G. Denisov, Y.V. Bykov, M.Yu. Glyavin, A.G. Luchinin, M.M. Morozkin, D.I. Sobolev. High efficient gyrotron-based systems for technological applications. Proc. 33rd Int. Conf. on Infrared, Millimetre and Terahertz Waves, Pasadena, CA, USA, 1-2, 2008.

[Edg93] Ch. J. Edgcombe, editor. Gyrotron Oscillators: their principles and practice. Taylor & Francis, Washington, DC, 1993.

[Fal77] L.R. Falce. Iridium cathode development. NRL Contract N00173-76-C-0296, Final Report, April 1977.

[Fal89] L.R. Falce. A cavity reservoir dispenser cathode for CRTs and low cost traveling wave tubes. IEEE Trans. Electron Devices, vol. 36, no. 1, pp. 169–172, Jan. 1989.

[Fal90] L.R. Falce. Thermionic cathodes: a review. Presented at the Proc. Workshop on Short Pulse High Current Cathodes, Bendor, France, 1990.

[Fal91] L.R. Falce. Controlled porosity dispenser cathodes for high resolution CRTs. In 1991 Soc. Information Display Int. Symp. Dig., 1991.

[Fal92] L.R. Falce. Dispenser cathodes go commercial. Presented at the Proc. Tri-Service/NASA Cathode Workshop, 1992.

[Fal92] L.R. Falce. CRT dispenser cathodes using molybdenum rhenium emitter surfaces. Presented at the Proc. Society Inf. Display Int. Symp. Dig., 1992.

[FB86] L.R. Falce and G.S. Breeze. Controlled porosity dispenser cathode. U.S. Patent 4 587 455, May 6, 1986.

[FGPY77] V.A. Flyagin, A.V Gaponov,M.I. Petelin, and Y.K. Yulpatov. The Gyrotron. IEEE Transactions on Microwave Theory and Techniques, MTT-25:514–521, 1977.

[Fla12] Jens H. Flamm. Diffraction and Scattering in Launchers of Quasi-Optical Mode Converters for Gyrotrons. PhD thesis, Karlsruhe Institute of Technology, Karlsruhe, KIT Scientific Publishing, ISBN 978-3-86644-822-3, 2012.

[FPA+15] J. Franck, I.G. Pagonakis, K.A. Avramidis, G. Gantenbein, S. Illy, M. Thumm, and J. Jelonnek. Magnetron injection gun for a 238 GHz 2 MW coaxial-cavity gyrotron. Microwave Conference

GeMiC 2015, March 16–18, 2015, Nürnberg, Germany, pages 260-263.

[FSJ67] A.I. Figner, A.I. Soloveickik, and I.V. Judinskaja. US Patent 3 358 178, Dec. 1967.

[Gap59] A.V. Gaponov. Interaction Between Electron Fluxes and Electromagnetic Waves in Waveguides. Izv. Vyssh. Uchebn. Zahved. Radiofizika, Vol. 2, pages 450–462, 1959.

[GB04] G. Graertner and D. Barratt. Life-limiting mechanisms in Ba-oxide, Ba-dispenser and Ba-Scandate cathodes. In Proc. 5th Int. Vac. Electron Sources Conf., Sep. 2004, pages 59–61.

[GDGT89] C.E. Garner, W.D. Deininger, J. Gibson, and R. Thomas. Patterning of dispenser cathode surfaces to a controlled porosity. IEEE Trans. Electron Devices, Vol. 36, No. 1, pages 158–168, Jan. 1989.

[GF02] T.J. Grant and L.R. Falce. Life predictions model for vacuum electron devices using barium calcium aluminate impregnated tungsten cathodes as electron sources. In Proc. IEEE Int. Vacuum Electronics Conf., Apr. 2002, pages 300–307.

[GGG+65] A.V. Gaponov, A.L. Goldenberg, D.P.Grigoryev, I.M. Orlova, T.B. Pankratova, and M.I. Petelin. 1965. Induced synchrotron radiation of electrons in cavity resonators. Pis'ma JETP, 2: 430-435, (JETP Letters, 2: 267-269).

[Gil86] A.S. Gilmour. Microwave Tubes. Copyright 1986, Artech House, Inc. Dedham, MA 02026, ISBN: 0-89006-181-5.

[HCR51] R.C. Hughes, P.P. Coppola, and E.S. Rittner. US Patent 2 700 118, Nov. 1951.

[Hig08] High Current Density Cathodes for Future Vacuum Electronics Applications, Final Report, U.S. Air Force Office of Scientific Research contract FA9550-07-C-0063, Apr. 2008.

[Hou67] J.M. Houston. High work-function coating on barium dispenser cathodes. GE Report 67-C-223, June 1967.

[HW51] I.G. Hermann, and P.S. Wagener. The oxide-coated cathodes. Chapman and Hall Ltd., 1951.

[IB99] S. Illy and E. Borie. Investigation of beam instabilities in gyrotron oscillators using kinetic theory and particle-in-cell simulation. Journal of Plasma Physics, 62(01): pages 95–115, 1999.

[IFSW05] R.L. Ives, L.R. Falce, S. Schwartzkopf, and R. Witherspoon, Controlled Porosity Cathodes from Sintered Tungsten Wires, IEEE Trans. on Electron Devices, Vol. 52, No. 12, pp. 2800-2805, December 2005.

[Ive11] R.L. Ives. private communication, 2011.

[Ive12] R.L. Ives. private communication, 2012.

[JAF+13] J. Jelonnek, K. A. Avramidis, J. Franck, G. Gantenbein, K. Hesch, S. Illy, J. Jin, A. Malygin, I. Gr. Pagonakis, T. Rzesnicki, A. Samartsev, T. Scherer, A. Schlaich, M. Schmid, D. Strauss, J. Zhang, and M. Thumm. KIT gyrotron development for future fusion applications, [Mo10-3]. In 38th International Conference on Infrared, Millimeter, and Terahertz Waves (IRMMW-THz), Mainz, Germany, Sept 2013.

[JHZ+04] D. Jiang, S. Hong, C. Zhou, D. Wang, and X. Liu. Preparation of impregnated barium Scandate cathode and its application. In

Proc. 5th Int. Vac. Electron Sources Conf., Sep. 2004, pages 206–207.

[JLL00] K. L. Jensen, Y. Y. Lau, B. Levush. Migration and escape of barium atoms in a thermionic cathode. IEEE Transactions on Plasma Science, Vol. 28, No. 3, June 2000, Pages 772 - 781.

[Jon60] W.D. Jones. Principles of Powder Metallurgy. London, U.K.: Edward Arnold, 1960.

[JPT+06] J. Jin, B. Piosczyk, M. Thumm, T. Rzesnicki, and S. Zhang. Quasi-optical mode converter/mirror system for a high power coaxial-cavity gyrotron. IEEE Transactions on Plasma Science, Vol. 34, No. 4, pp-1508-1515, 2006.

[KBT04] M. V. Kartikeyan, E. Borie, and M. Thumm. Gyrotrons: high power microwave and millimetre wave technology. Advanced texts in physics. Springer, Berlin, 2004.

[Ker96] S. Kern. Numerische Simulation der Gyrotron Wechselwirkung in koaxialen Resonatoren. Wissenschaftliche Berichte des Forschungszentrums Karlsruhe FZKA 5837, PhD thesis, Universität Karlsruhe (TH), 1996. German.

[Lev53] R. Levi. New dispenser type thermionic cathode. Appl. Phys., 1953, 24, p.233.

[Lev55] R. Levi. Improved impregnated cathode. ibid., 1955, 26, p. 639.

[LFMC10] R. Lawrence Ives, L.R. Falce, G.V. Miram, and G.A. Collins. Controlled porosity cathodes for high current density applications. IEEE Trans. Plasma Sci., Vol. 38, No. 6, pages 1345–1353, Jun. 2010.

[LFMC12] R. Lawrence Ives, L.R. Falce, G.V. Miram, and G.A. Collins. Advanced Dispenser Cathodes. IEEE Trans. Plasma Sci., Vol. 40, No. 5, pages 1299–1302, May 2012.

[LFT+99] G. Link, L. Feher, M. Thumm, H. -J. Ritzhaupt-Kleissl, R. Bohme and A. Weisenburger. Sintering of advanced ceramics using a 30-GHz, 10-kW, CW industrial gyrotron. IEEE Transactions on Plasma Science, Vol. 27, Issue 2, Pages 547 - 554, 1999.

[LHJ+04] G. Link, M. Hauser-Fuhlberg, M. Janek, R. Nuesch, S. Takayama, M. Thumm, A. Weisenburger. Millimetre-wave high temperature processing of powders. Joint 29th International Conference on Infrared and Millimetre Waves and 12th International Conference on Terahertz Electronics, Pages: 829 - 830, 2004.

[LH52] R. Levi and R.C. Hughes. US Patent 2 700 000, Feb. 1952.

[Lie10] C. Lievin. Coaxial Gyrotron 170GHz/2MW. Conference F4E. October 20th 2010.

[LJL50] H.J. Lemmens, M.J. Jansen, and R. Loosjes. A new thermionic cathode for heavy loads, Philips Tech. Rev., 1950, 11, pages 341-350.

[LZW+04] W. Liu, K. Zhang, Y. Wang, K. Pan, X. Gu, J. Wang, J. Li, and M. Zhou. Operating model for Scandate cathode with scandia doped tungsten bodies. In Proc. 5th Int. Vac. Electron Sources Conf., Sep. 2004, pages 62–64.

[Man06] V.N. Manuilov, "Numerical simulation of low-frequency oscillations of the space charge and potential in the electron-optical system of a gyrotron". Radiophysics and Quantum Electronics, Vol. 49, No. 10, 2006, pages 786-792.

[MGDL08] M.V. Morozkin, M.Y. Glyavin, G.G. Denisov, A.G. Luchinin, "A high-efficiency second-harmonic gyrotron with a depressed collector", International Journal of Infrared and Millimeter Waves 29 (11), 1004-1010, 2008.

[MP09] V.N. Manuilov, S.A.Polushkina, "Behavior of helical electron beams in gyrotrons with high pitch factors", Radiophysics and Quantum Electronics, Vol. 52, No. 10, pages 714-721, 2009.

[MZZ+10] A.V. Malygin, N.I. Zaitsev, S.A. Zapevalov, M.A. Moiseev, A.S. Shevchenko. Studying the start-up scenario for a pulsed gyrotron with a relativistic electron beam. Radiophysics and Quantum Electronics, 2010, Volume 53, Issue 3, Pages: 178-181.

[NTP14] G. S. Nusinovich, M. Thumm, and M. Petelin. The gyrotron at 50: Historical overview. Journal of Infrared, Millimeter, and Terahertz Waves, 35(4): pages 325–381, 2014.

[Nus04] G.S. Nusinovich. Introduction to the physics of gyrotrons. John Hopkins studies in applied physics. Johns Hopkins Univ. Press, Baltimore, 2004.

[Pio12] B. Piosczyk. Private communication, 2012.

[Pio93] B. Piosczyk. Electron guns for gyrotron applications. Edgcombe, C.J. (Hrsg.). Gyrotron Oscillators - Their Principles and Practice, Chapter 5, London: Taylor and Francis, pp 123-146, 1993.

[PV04] J.Gr. Pagonakis and J.L. Vomvoridis. The self-consistent 3D trajectory electrostatic code Ariadne for gyrotron beam tunnel simulation. Joint 29th Int. Conf. Infrared Millimiter Waves and 12th Int. Conf. THz. Electronics, 2004, pp. 657-658, Karlsruhe, Germany.

[RDKB06] M. Ravi, P. Devi, K. Kumar, and K. Bhat. Characterization of W–Ir mixed metal matrix Scandate cathode. In Proc. IEEE Int. Vac. Electron. Conf., Monterey, CA, Apr. 2006, pages 57–58.

[RCY02] H.C. Rho, S.L. Cho, and H.T. Yang. Cathode for electron tube and method of preparing the same. U.S. Patent 2002/0 041 141A1, Apr. 11, 2002.

[RKB11] M. Ravi, K.S. Kumar, K.S. Bhat. Tungsten-rhenium mixed metal matrix cathodes. In Proc. 2011 IEEE International Vacuum Electronics Conference (IVEC), Pages: 39 – 40, Bangalore, India.

[RLFT00] S. Rhee, G. Link, L. Feher, M. Thumm. High power millimetre waves for sintering of nanostructured ceramics. 25th International Conference on Infrared and Millimetre Waves, Pages: 425 – 426, 2000.

[Sch15] A. Schlaich. Time-dependent spectrum analysis of high power gyrotrons. Karlsruher Forschungsberichte aus dem Institut für Hochleistungsimpuls- und Mikrowellentechnik, Band 6, PhD thesis, Karlsruhe Institute of Technology (KIT), 2015. ISBN 978-3-7315-0375-0.

[SBC+11] E.A. Soluyanova, Yu.V. Bykov, A.V. Chirkov, G.G. Denisov, E.A. Kopelovich, V.B. Orlov, A.B. Pavelyev, E.V. Sokolov, E.M. Tai. Gyrotron complexes for technological applications. Proc. 8th Int. Workshop Strong Microwaves and Terahertz Waves: Sources and Applications, Nizhny Novgorod – St. Petersburg, Russia, July 9-16, p. 133, 2011.

[Sch59] J. Schneider. Stimulated Emission of Radiation by Relativistic Electrons in a Magnetic Field. Physical Review Letters, Vol. 2: pages 504–505, 1959.

[SCL+06] G. Schietrum, G. Caryotakis, N. Luhmann, Y. Wang, J. Li, G. Miram, and B. Vancil. 100 A/cm^2 tungsten-scandate nanopowder thermionic cathode testing at SLAC. In Proc. Multidisciplinary Univ. Res. Initiative Annu. Rev., Monterey, CA, Apr. 2006.

[Shu04] W. Shuguang. Scandate cathode for TWT. In Proc. 5th Int. Vac. Electron Sources Conf., Sep. 2004, pages 224–225.

[Sli65] M.J. Slivka. Investigation of high emission density cathode. USAFML Technical Report 65-196, June 1965.

[SYH04] A. Shih, J. E. Yater, and C. Hor. Scandate cathode fundamentals: Interactions of Ba, O, and Sc on W. In Proc. 5th Int. Vac. Electron Sources Conf. Sep. 2004, pages 57–58.

[TG83] R.E. Thomas and R.F. Green. Controlled porosity sheet for thermionic dispense cathode and method of manufacture. U.S. Patent 4 379 979, Apr. 12, 1983.

[Thu15] M. Thumm. State-of-the art of high power gyro-devices and free electron masers. KIT Scientific Report 7693, Karlsruhe Institute of Technology (KIT), Karlsruhe, 2015.

[TLST04] S. Takayama, G. Link, M. Sato, M. Thumm. Sintering of metal powder samples with millimetre wave technology. Joint 29th International Conference on Infrared and Millimetre Waves and 12th International Conference on Terahertz Electronics, Pages: 729 – 730, 2004.

[Tuc78] R.A. Tuck. Surface studies by methods which are unique to thermionic emitters. Tri-Service cathode workshop, Washington DC, Jan.-Feb. 1978.

[Twi58] R.Q. Twiss. Radiation Transfer and the Possibility of Negative Absorption in Radio Astronomy. Australian Journal of Physics, Vol. 11: pages 567–579, 1958.

[VJK09] J.M. Vaughn, K.D. Jamison, and M.E. Kordesch. An in situ emission microscopy study of barium supply to the surface of scandium-coated porous tungsten. In Proc. IEEE Int. Vac. Electron. Conf., Rome, Italy, Apr. 2009, pages 559–560.

[VK71] A.J.A. Van Stratum and P.N. Kuin. Tracer study on the decrease of the emission density of osmium-coated impregnated cathodes. Appl. Phys., 1971, 42, pages 4436-4437.

[VVBZ77] A.J.A. Van Stratum, J.G. Van Os, J.R. Blatter, and P. Zalm. US Patent 4 007 393, Feb. 1977.

[Wij92] J.F.C.M. Wijen. Impregnated cathodes with a controlled porosity. U.S. Patent 5 118 317, Jun. 1992.

[WLG+09] J. Wang, W. Liu, T. Gao, L. Li, and M. Zhou. Emission properties of impregnated cathodes with Y2O3 co-doped tungsten substrate. In Proc. IEEE Int. Vac. Electron. Conf., Rome, Italy, Apr. 2009, pages 525–526.

[YGP+05] H. Yuan, X. Gu, K. Pan, Y. Wang, W. Liu, K. Zhang, J. Wang, M. Zhou, and J. Li. Characteristics of Scandate-impregnated cathodes with sub-micron scandia-doped matrices. Appl. Surf. Sci., vol. 251, no. 1–4, pages 106–113, Sep. 2005.

[ZV66] P. Zalm, and A.J.A. Van Stratum. Osmium dispenser cathodes. Philips Tech. Rev., 1966, 27, pages 69-75.

[ZZLL04] M. Zhang, H. Zhang, P. Liu, and Y. Li. Investigation of recovery characteristics after ion bombardment on Scandate cathode. In Proc. 5th Int. Vac. Electron Sources Conf., Sep. 2004, pages 153–155.

[ZZMS11] N. A. Zavol'skiy, V.E. Zapevalov, M.A. Moiseev and A.S. Sedov. Influence of the axial misalignment of the electron beam and the cavity on the gyrotron parameters. Radiophysics and Quantum Electronics, Vol. 54, No. 6, November, 2011 (Russian Original Vol. 54, No. 6, June, 2011).

A Appendix

A.1 CPR cathode

In Section 1.5.3 the general structure and manufacturing procedures of CPR cathodes have been discussed. Here in Appendix A1 the design of the segmented CPR emitter suitable for the gyrotron designed within this dissertation will be shown. The technical design of the CPR cathode configuration has been done by CCR.

The configuration of the gyrotron electron gun is shown in Fig. A.1. The modular design of the electron gun allows to exchange different types of emitters. In Fig. A.2 the CPR emitter configuration is shown which consists of twelve segments separated by thin ceramic insulators. Figure A.3 shows the drawing of one CPR emitter segment, which consists of heater, barium reservoir and tungsten wire emitter. A photo of the real segment assembled by CCR is shown in Fig. A.4. When heated to operating temperature, the work function reducing material diffuses through the pores to the emission surface. For thermionic dispenser cathodes, the work function reducing material is usually a mixture of barium oxide, calcium oxide, and aluminum oxide.

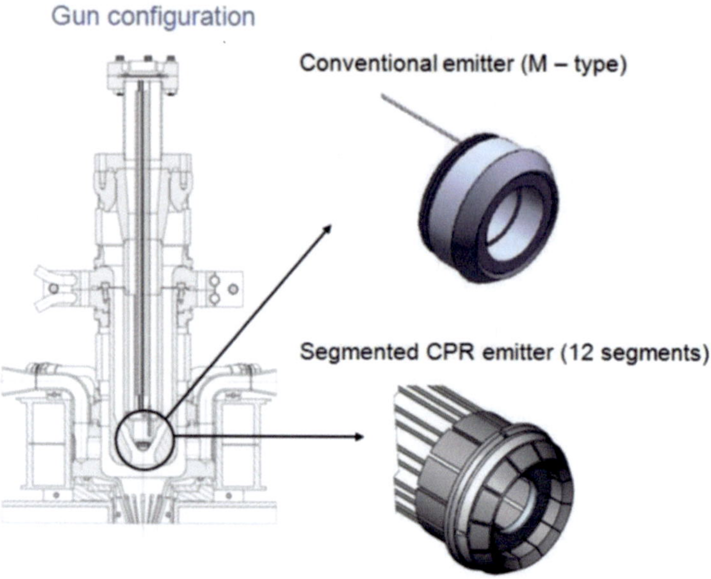

Figure A.1: Gun configuration with segmented CPR emitter

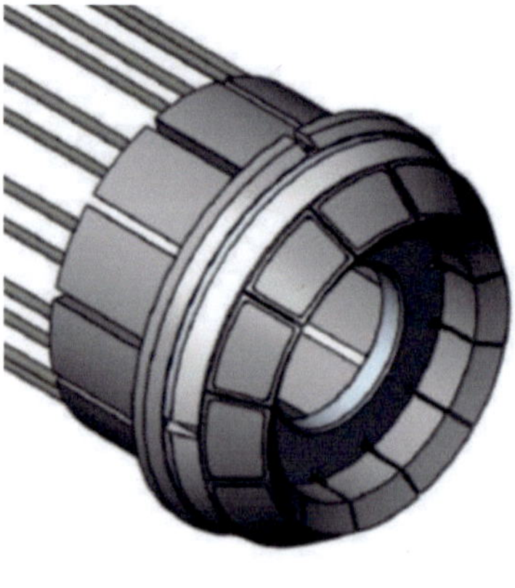

Figure A.2: Segmented CPR emitter configuration (12 segments)

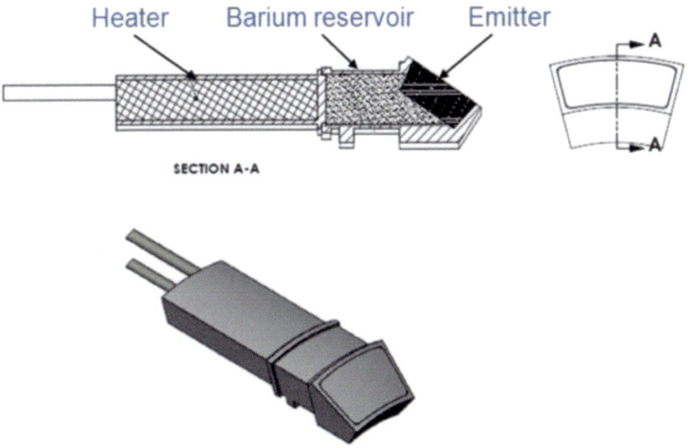

Figure A.3: Configuration of a CPR emitter segment

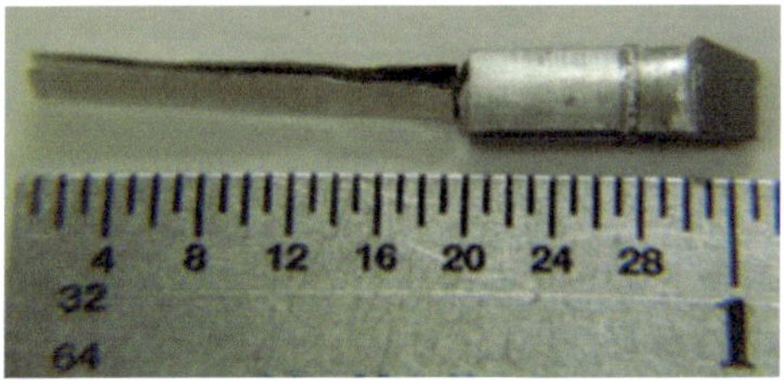

Figure A.4: Photo of one segment of the CPR emitter

A.2 Glidcop material properties

Glidcop is the registered trade name of SCM Metal Products for copper which is dispersion-strengthened with ultra-fine particles of aluminium oxide. Table A.1 lists material properties of different Glidcop types and includes the data for oxygen-free high conductivity copper (OFHC). All the data is given for room temperature 20 °C.

Table A.1: Material properties for various type of Glidcop

	OFHC	AL-15	AL-25	AL-60
Al_2O_3 content [wt.%]	0	0.7	1.2	2.7
Density [g/cm^3]	8.94	8.9	8.86	8.81
Melting Point [°C]	1083	1083	1083	1083
El. conductivity [MS/m]	58	54	50	45
Isotr. therm. conductivity [W/ °Cm]	391	365	344	322
Coeff. therm. expansion [µm/m°C]	17.7	16.6	16.6	16.6
Young's Modulus [GPa]	115	130	130	130
Specific heat [J/kg°C]	38.5	38.5	38.5	38.5
Ultimate tensile strength [MPa]	220	360	390	470

Acknowledgements

The present work was accomplished during my time as research assistant at the Institute for Pulsed Power and Microwave Technology (IHM) of the Karlsruhe Institute of Technology (KIT).

Firstly, I would like to express my sincere gratitude to my doctoral advisor Prof. Dr. rer. nat. Dr. h.c. Manfred Thumm for the continuous support of my Ph.D. study and related research, for his patience, motivation, and immense knowledge. His guidance helped me during all the time of research and writing of this thesis. I could not have imagined having a better advisor and mentor for my Ph.D. study.

I am grateful to Prof. Dr.-Ing. Arne F. Jacob from TU Hamburg-Harburg for his great interest in my work and being my second reviewer. I also would like to thank the Director of IHM Prof. Dr.-Ing. John Jelonnek for his kind guidance, continuous support and fruitful discussions.

I am deeply grateful to Prof. Dr. Edith Borie for the meticulus and productive proofreading she provided for the thesis.

I would like to thank all members of the High Power Microwave Department at IHM for the excellent working environment, valuable discussions and continuous support in the course of this project. Especially, I would like to thank Dr. Stefan Illy for the innumerable discussions, constant encouragement and support, without his help this work would not have been possible. I wish to express my deep gratitude and sincere thanks to Dr. Konstantinos A. Avramidis and Dr. Ioannis Pagonakis for their help and valuable discussions throughout this work. I would also like to acknowledge Dr. Stefan Kern, Dr. Gerd Gantenbein, Dr. Bernhard Piosczyk, Dr. Günter Dammertz and Dr. Guido Link for valuable technical advices and fruitful discussions at the beginning of this project.

It is my pleasure to acknowledge the excellent work of IHM drafter Jörg Weggen who did the detailed drawings of the triode electron gun, the beam tunnel, the collector, the gyrotron test facility and the water cooling system. My sincere thanks also go to the Calabazas Creek Research Company in the USA, in particular to Dr. Lawrence Ives, David Marsden and George Collins, who helped in the design of the electron gun, collector and output window and manufactured the gyrotron cathode and collector. I am grateful to IHM engineers Daniel Mellein, Daniel Papenfuß, Wolfgang Leonhardt, Martin Schmid, Klaus Baumann and the Ph.D. students Chuanren Wu and Andreas Schlaich for collaboration during design and installation of the gyrotron control system. My sincere thanks also go to IHM technicians Rouven Lang and Walter Spieß for their work and support during assembly of the gyrotron test stand. I am extremely thankful to the Ph.D. students Sebastian Ruess, Joachim Franck, Jianghua Zhang and Parth Kalaria for their help during the experimental investigation of the gyrotron.

Very special thanks go to my family: my parents and my brother, for their unconditional support and encouragement throughout my work and my life in general.

KARLSRUHER FORSCHUNGSBERICHTE AUS DEM INSTITUT FÜR HOCHLEISTUNGSIMPULS- UND MIKROWELLENTECHNIK
(ISSN 2192-2764)

Die Bände sind unter www.ksp.kit.edu als PDF frei verfügbar oder als Druckausgabe bestellbar.

Band 1 **MATTHIAS BERINGER**
Design Studies towards a 4 MW 170 GHz Coaxial-Cavity Gyrotron. 2011
ISBN 978-3-86644-663-2

Band 2 **JENS FLAMM**
Diffraction and Scattering in Launchers of
Quasi-Optical Mode Converters for Gyrotrons. 2012
ISBN 978-3-86644-822-3

Band 3 **MATTIA DEL GIACCO**
Investigation of Fretting Wear of Cladding Materials in Liquid Lead. 2013
ISBN 978-3-86644-960-2

Band 4 **AMITAVO ROY CHOUDHURY**
Investigations of After Cavity Interaction in Gyrotrons
Including the Effect of Non-uniform Magnetic Field. 2013
ISBN 978-3-7315-0129-9

Band 5 **MICHAEL BETZ**
The CERN Resonant WISP Search (CROWS). 2014
ISBN 978-3-7315-0199-2

Band 6 **ANDREAS SCHLAICH**
Time-dependent spectrum analysis of high power gyrotrons. 2015
ISBN 978-3-7315 0375 0

Band 7 **DHIDIK PRASTIYANTO**
Temperature- and Time-Dependent Dielectric Measurements
and Modelling on Curing of Polymer Composites. 2016
ISBN 978-3-7315-0424-5

Band 8 **YIMING SUN**
Adaptive and Intelligent Temperature Control of Microwave
Heating Systems with Multiple Sources. 2016
ISBN 978-3-7315-0467-2

Band 9 **JIANGHUA ZHANG**
Influence of Emitter surface roughness and Emission inhomogeneity
on Efficiency and stability of high power Fusion gyrotrons. 2016
ISBN 978-3-7315-0578-5

KARLSRUHER FORSCHUNGSBERICHTE AUS DEM
INSTITUT FÜR HOCHLEISTUNGSIMPULS- UND MIKROWELLENTECHNIK
(ISSN 2192-2764)

Die Bände sind unter www.ksp.kit.edu als PDF frei verfügbar oder als Druckausgabe bestellbar.

Band 10 **ANTON MALYGIN**
Design and Experimental Investigation of a Second Harmonic 20 kW Class
28 GHz Gyrotron for Evaluation of New Emitter Technologies. 2016
ISBN 978-3-7315-0584-6